W9-CPV-126

Praise for *Wireless Wars*

"*Wireless Wars* captures the totality of the decades-long commercial and government effort by the CCP to dominate telecommunications infrastructure and operations across the entire globe. By avoiding technical jargon and highlighting key concepts through pithy summaries of hundreds of insider interviews, *Wireless Wars* translates a complex subject into a clear, articulable framework on how we got into this predicament and how we can get out of it. Jon Pelson's thirty-plus years of in-depth experience within the US telecommunications manufacturing and operations sector fused with well-researched and meticulously documented source material on Huawei, ZTE, and other PRC telecommunications companies makes for an extremely compelling must-read for industry experts, intelligence professionals, policy makers, and concerned citizens interested in protecting our nation."

—William Evanina, former director of the
National Counterintelligence Center

"*Wireless Wars* shines a bright light on the ongoing battle for national supremacy and global security in cyberspace, with China and the United States being the primary actors, and 5G networks providing the latest battleground. Jon Pelson has an insider's knowledge of this territory, and his book has important implications not only for policy makers, defense leaders, and technology investors, but for interested citizens, as well."

—Geoffrey A. Moore, international bestselling author of *Crossing the Chasm*

"In *Wireless Wars*, Jon Pelson accurately depicts the transfer, both consciously and unconsciously, of US wireless equipment manufacturing to China. Having been at some of those meetings he depicts in the late 1990s and early 2000s and scratching my head, asking, 'What's really going on here?', he fills in a lot of blanks that now, in retrospect, make sense. He touches on some important points about how seemingly benign behaviors, such as increased standards participation, global manufacturing, and investment policy, have turned out to have sinister motivations and results. As he notes, this is not just about getting cheaper network infrastructure. This is about what happens when a hostile government subsidizes vendors to take over a market, force the bankruptcy or mergers of vendors who played by the rules and had millions of dollars of R&D stolen, and the resulting national security threats of having untrusted vendors subject to government control with their equipment sitting in some of our most sensitive networks."

—Eric Burger, former chief technology officer of the Federal
Communications Commission and professor at Georgetown University

"*Wireless Wars* chronicles how America squandered a virtually insurmountable technological lead in the telecommunications equipment market, allowing China to gain preeminence. It makes a compelling argument on the necessity of the US regaining leadership in this vital sector. More important, it charts a path for achieving that goal. A must-read for anyone interested in 5G, innovation, and national competitiveness."

—Joseph B. Fuller, professor of management practice at
Harvard Business School and cofounder of Monitor Group

"China's rulers are waging a war for global dominance on many fronts—the technological battlefield prominent among them. In *Wireless Wars,* Jonathan Pelson brings clarity to how we got here, how high the stakes are, and how we can win."

—Clifford D. May, president of the Foundation for Defense of Democracies

WIRELESS
WARS

WIRELESS
WARS

CHINA'S DANGEROUS DOMINATION OF 5G AND HOW WE'RE FIGHTING BACK

JONATHAN PELSON

BenBella Books, Inc.
Dallas, TX

Wireless Wars copyright © 2021 by Jonathan Pelson

All rights reserved. No part of this book may be used or reproduced in any manner whatsoever without written permission of the publisher, except in the case of brief quotations embodied in critical articles or reviews.

BenBella Books, Inc.
10440 N. Central Expressway
Suite 800
Dallas, TX 75231
www.benbellabooks.com
Send feedback to feedback@benbellabooks.com

BenBella is a federally registered trademark.

Printed in the United States of America
10 9 8 7 6 5 4 3 2 1

Library of Congress Control Number: 2021020938
ISBN 978-1-953295-61-3 (trade cloth)
ISBN 978-1-953295-95-8 (electronic)

Editing by Vy Tran
Copyediting by Scott Calamar
Proofreading by Kim Broderick and Cape Cod Compositors, Inc.
Indexing by Debra Bowman
Text design by PerfecType, Nashville, TN
Cover design by Pete Garceau
Cover image © iStock / Dmitry Volkov (wifi circles) and Tania Bondar (countries)
Printed by Lake Book Manufacturing

Special discounts for bulk sales are available. Please contact bulkorders@benbellabooks.com.

To my father, who taught me how to write.

CONTENTS

PROLOGUE

When the COVID-19 virus first emerged in December of 2019, the Chinese government sprang into action, deploying a suite of world-class technology to keep people safe and prevent the spread of the disease. Within weeks of the first cases, mobile smartphone applications had been created and pushed out to protect the country's 1.4 billion people, using China's superior wireless data networks integrated with publicly deployed body-temperature sensors, facial-recognition devices, and building-security controls. These systems were tied together using artificial intelligence technology that made sense of the massive flood of information and enabled authorities to impose restrictions on end users while notifying police of lapses in citizens' behavior.

These mobile-based controls tracked people's locations by using the GPS feature of their phones to aid in contact tracing and determine when they were in close proximity to others who were at high risk of infection. Cash payment was banned in favor of touchless (and trackable) electronic transactions, allowing authorities to monitor health supply purchases at drugstores or supermarkets, whether made online or in person. The information was used to generate health scores for each person and to block citizens from entering buildings or leaving their own homes if they scored too low.

It was a masterful use of mobile technology to serve the public good, though some observers were puzzled at how quickly the country was able to pull together such a comprehensive suite of monitoring and tracking applications. In fact, it appears that China had been forced to make public the full capabilities of the technology-driven surveillance state that the Chinese Communist Party (CCP) had spent the past several years putting in place: deploying a combination of mobile technology, public sensors, and artificial intelligence to observe and control everything its citizens said, did, or bought, even who they met and communicated with. The result, in this case, was an abrupt reduction in the spread of the disease. Even Wuhan, the city at ground zero of the deadly pandemic, escaped the worst consequences felt by the rest of the world.

Globally, a shortage of personal protective equipment exacerbated the crisis, and medical centers around the world discovered that many of their supplies came exclusively from Chinese factories, and they were arriving with serious quality problems—or not at all. It appeared that China, not interested in experiencing its own internal shortages, or simply flexing its muscle over trade rivals, wasn't shipping on its contracted agreements.

Yet on January 30, 2020, the World Health Organization (WHO) issued a press release on the novel coronavirus praising "China's commitment to transparency and to supporting other countries." By the time of the release, internal emails and transcripts of meetings were already circulating at the WHO that revealed a very different assessment. The Associated Press, in its investigation of the WHO, found that China had delayed notifying global health authorities of the details of the disease for weeks, hastening the spread of the illness from Europe to North America and then to the rest of the world.

As the initial wave of the crisis passed, global leaders started asking questions about what took place in the first months of the pandemic: how China shared what it knew and whether it had lied to other countries' health authorities. But when Australia's prime minister announced a desire for an inquiry, the Chinese government responded with across-the-board trade sanctions, breaking international treaties and letting the world know not to push it on the matter.

Seeing the belligerence from the CCP and worrying about the vulnerability that came from using China as a single-source supplier of vital goods, corporate and government leaders around the world began to reconsider another critical technology that was being supplied, increasingly, from a single Chinese vendor: the very mobile telephone network gear that had enabled China's rapid response to the pandemic. China's technology giant Huawei, the $120 billion company that had delivered so much of the surveillance and monitoring capability used to protect the country in the early days of the crisis, had taken a massive lead in the development and sale of wireless equipment and was poised to dominate the world market for the new networks being deployed for 5G mobile services. This was the technology that would enable driverless cars, automated homes, and connected factories—the Internet of Things. Solutions that could make a country more productive or protect against the next pandemic.

Concerns had been raised in recent years that Huawei's equipment presented a security risk, as claims arose that China had used Huawei's gear to spy on other countries, and this new worry increased the apprehension. As some countries expressed reservations about deploying Huawei equipment into their networks, China threatened retaliation, warning Germany, for example, that the future of its automotive markets in China could be jeopardized by a German ban on Huawei. Similar warnings went out to other countries.

At the same time, wireless service operators around the world realized that, perhaps, they didn't have many choices anyway. The Western companies that had invented mobile telephony were nearly all bankrupt, driven out by the low prices and strong technology of the Chinese giants. For years, companies and consumers had been seduced by low prices and shiny technology coming from a country that was now looking less like a trading partner and more like a geopolitical rival, asserting its might through the use of trade dependence, technology domination, and financial muscle.

But if the world couldn't turn to China for this critical 5G technology, what *was* the alternative?

INTRODUCTION

Working in Beijing in the late nineties, I found myself unable to bribe my customers.

I was selling Lucent Technologies' wireless communications gear to companies all over the world, including the large Chinese phone companies, and my competitors were taking business away through their liberal use of inducements. Cars and cash, mainly. The problem was, I didn't work that way, my company didn't tolerate such behavior, and my government (unlike many other countries') considered bribery a crime, under the Foreign Corrupt Practices Act. Yet there I was, trying to compete with other companies that had no such compunctions, chasing customers all too willing to put themselves in play.

So the buyers wouldn't give our salespeople the time of day, and they didn't always seem interested in making a purchasing decision based on product quality or price anyway. In the end, I developed a solution that shot two vultures with one arrow, as our hosts might say: Lucent flew in professors from Harvard Business School to teach classes in supply-chain management to executives from the Chinese carrier community. We set up seminars (translated into Mandarin, of course) on Western business practices to equip a phone company with gear that would achieve their mission. The telecom execs learned something useful, hopefully recognizing the cost to their companies of throwing hundreds of millions of dollars in business to whoever parked a Volvo in their driveway, and they left our training sessions carrying a personalized achievement certificate bearing the Harvard Business School logo. For a culture that values brand almost as much as it reveres education, this had enormous appeal. Everyone who attended the sessions felt good about Lucent for sponsoring the free seminars, and we were able to deal with more educated buyers. My company established a huge business in China, taking a lead role in the largest telecommunications build-out in history.

At the time, North American and European telecommunications equipment companies were so far ahead of China that we thought nothing of educating Chinese customers, partners, and even competitors on how to do things better. Yet, for at least some of the Chinese executives, those business classes launched them on the road to world domination in the telecom sector. We taught them how to cook food Western style, and now they're eating our lunch.

$$\bullet \ \bullet \ \bullet$$

A few years ago, not many people had heard of Huawei (pronounced "wahway"), at least in the United States, and little thought was given to 5G—fifth-generation mobile telephone networks—beyond the commercials claiming that it was "a hundred times faster than 4G." But Huawei has grown from a $5,000 seed investment in 1987 into a $120 billion company—bigger than BMW, General Electric, or Boeing, and bigger than all the other major telecom network equipment makers in the world combined, while also surpassing Samsung and Apple as the world's largest maker of smartphones. Their twenty-billion-dollar-a-year research-and-development budget is larger than most competitors' entire revenue streams.

By 2019 it looked like China's most successful international company was going to be building the world's 5G networks; they made the cheapest equipment, and it also happened to be the best. They had dominant market share in the industry and were the lead supplier in Europe, Asia, Africa, and Latin America.

But suspicions had been growing in the intelligence services of the "Five Eyes" nations—the United States, Canada, the United Kingdom, Australia, and New Zealand—and the FBI set up a team charged with determining what Huawei was after. They studied the company's failed efforts to get a foothold among the Tier 1 service providers in the US market, with sales limited to smaller, regional, and rural cellular companies, a far cry from the Sprint and AT&T accounts they really wanted. Even those sales were won only by practically giving the equipment away. Why would Huawei even bother?

The answer became clear as a pattern emerged; according to counter-intelligence agents at the FBI, the map of their rural network deployment overlaid too closely with America's most secure defense facilities: Air Force Global Strike Command, United States Special Operations Command (SOCOM), and intercontinental ballistic missile (ICBM) bases. Far too many of them were being served by wireless equipment provided and managed by China. Other countries around the world also became concerned that the CCP might start using its national equipment giant to compromise foreign networks.

And then something happened that changed the way the world viewed China. Suddenly, in 2020, the worst pandemic since the Spanish Flu struck, originating in Wuhan, China. In their analysis of the COVID-19 crisis, countries began to pay closer attention to issues of Chinese government transparency, supply-chain vulnerability, and the dangers of relying on an authoritarian government with disturbing views on public disclosure and personal liberty.

Just when it seemed that China had reached a point where it might be the last man standing, controlling communications equipment for nearly all of the world, its national champion was brought to a grinding halt by a Western reaction it never expected. The unstoppable Huawei had its legs cut out by a shockingly aggressive US administration. Countries around the world began to reconsider their relationship with Huawei, surprising political observers who had seen the United States losing the support of its long-time allies. As one European technology leader confided to me, "We hate Trump, but secretly we love what he's doing about China."

These issues shouldn't have caused a panic about who was making the components that go into our communications networks—Chinese electronics are already pervasive—but there was an important difference between earlier wireless networks and the new 5G systems that were being rolled out. For the first time, the new generation of wireless technology was less about better cell phone services and more about transforming the way businesses and governments work. These new 5G networks access our financial information, control factories, deliver medical treatment, steer cars, and even link soldiers on the battlefield. Whoever builds these networks may be able to observe, throttle, or even terminate what passes over them. 5G is about a lot more than faster downloads, and the idea that a Chinese company with a questionable history would be responsible for delivering these services started to sound alarm bells in the halls of businesses, lawmakers, and intelligence agencies around the globe.

When the time rolled around to deploy 5G, governments across the world were torn by conflict. German chancellor Angela Merkel saw the first opposition she had ever faced from her own party when she tried to greenlight Huawei to deploy 5G across Germany, as the Bundestag pushed to ban the company. Other national leaders were pressed by their own parties to reverse course and even rip out the billions of dollars of Huawei gear that had been installed. And the world's two superpowers, the United States and China, marched to the brink of a full-fledged trade war, with the Democratic leaders in the US engaging in rare consensus with the Republican Trump administration over a complete ban of Chinese gear in American telecom networks.

• • •

How did we get to this point?

AT&T's Bell Laboratories invented cellular telephony in the sixties and seventies, competing with scientists at Motorola and Nortel, who together created one of the world's largest industries. Yet the top-tier North American and European manufacturers who quickly established themselves as standard-bearers blew a seemingly insurmountable lead in wireless communications by relocating factories to China and transferring technology to the local partners in order to get access to its cheap labor and its massive, growing customer market. These companies traded their future for short-term profits without considering the dangers of passing the lead to a country whose rulers have values that most people (Chinese citizens included) find reprehensible.

It was people, not companies, who made the decisions to relocate those factories to China, but they may not have been as foolish as they seem. Western telecom executives found themselves in a trap, faced with the choice between moving their manufacturing to low-cost locations (with the risk of an eventual loss of control) or keeping it local and ensuring their imminent defeat by any companies that *did* take advantage of cheap labor and R&D. It's easy to look back on the transfer of technology and manufacturing to China and say it was a boneheaded, short-sighted mistake. After all, there were plenty of boneheads in positions of authority at the Western equipment makers. But in many ways, their hands were forced by the circumstances described here.

Their mistake lay in failing to chart a "phase two" that would ensure *future* advancement of their business after moving their plants and technology to China. The Western companies had to acknowledge the risks, take advantage of the opportunity in the short term, and build a bridge from their boom of the nineties to twenty-first-century success. Instead, they emerged from the telecom meltdown in 2001 to find that the bridge was out and they were plunging into the abyss.

The rise of the Chinese telecom equipment giants like Huawei and ZTE (the state-owned rival to Huawei run by a former general in the People's Liberation Army) occurred exactly as the dot-com bubble burst, and with it the collapse of the service provider sector that comprised the customers of telecom equipment makers. Equipment sales fell in half, and service-provider consolidation led to a shift in power from the equipment makers to their customers. The result was the wipeout of North American and European equipment vendors. Once-magnificent organizations like Bell Labs fell into the hands of foreign acquirers, who also shortly found themselves collapsed and folded into *other* acquirers. How do you fit eight hundred pounds of potatoes into a sixty-pound sack? Take Lucent, Motorola, Nortel, and Alcatel, once worth nearly a trillion dollars between them, and roll them into Nokia and Ericsson, together valued at less than $60 billion in 2021.

This collapse signifies more than the loss of one industry sector. Telecom is not just critical infrastructure; it enables and controls all *other* critical infrastructures, from air travel to power grids to hospitals. Even the military, which already relies heavily on public communications networks, has become increasingly dependent on wireless public networks to execute its mission. Telecom networks, by their nature, are more vulnerable than other critical infrastructure because they allow a bad actor to cause trouble without needing to be physically on the scene or to breach a physical barrier. They are vulnerable to attacks from anyone, anywhere in the world, who is clever enough to access them. And they make other critical systems vulnerable to bad actors too.

• • •

The United States has lost leadership in technology areas in the past, with steel, automobiles, televisions. In most cases, the successors, like Japan and Korea, were based in countries that may have had very different cultures from America's, but they shared basic values of democracy and respect for human rights. All shared the concepts of representative government, beholden and accountable to the will of the people. It's not hard to imagine why the US wasn't supplanted by companies in authoritarian regimes; such systems rarely produce world-class commercial enterprises. So even as newly emergent countries competed against the United States and each other, it was a rivalry, not a war.

The battle for 5G brings a much different story. The Chinese Communist Party does not design or deploy telecom equipment. But the CCP demands that its national champions serve the national interest. With network equipment makers, this presents a grave danger: once a telecom system is deployed, the manufacturer is expected to remain active inside their customer's network, monitoring performance and managing and updating the equipment with security patches and performance enhancements.

Does China use this position to spy? Arguments can be made that "everybody does it." The United States has the best-funded spy apparatus in the world and uses it to listen in on friend and foe alike. But the similarities end there. Governments in democracies, imperfect as they may be, operate under different constraints and obligations than in China, where reports of suppression have emerged that exceed even the most dystopian fantasies of state control. Credible reports have emerged of China rounding up its own Muslim citizens in the far west provinces and placing them in brutal "re-education camps."

It's known that Chinese citizens are under constant scrutiny while in their home country. It's less known that many of them are unable to escape that scrutiny while abroad, even long after they have moved to a free society and adopted a new country and citizenship. For example, Chinese students studying at US universities are subject to punishment or arrest for views expressed in their classrooms, leading some top colleges to block student names from appearing on papers and exams.

There are those in Western intelligence communities who see China's handling of its own citizens as a live fire drill for how it could monitor, observe, and control rivals in the United States and elsewhere. The Chinese government can extend such abuses using its access to the world's communications networks, whether through databases believed to be hacked by the People's Liberation Army's infamous Advanced Persistent Threat unit or through eavesdropping, phishing, or socially engineered breaches into the most secure systems.

Huawei's role in the supply chain and the authorized access that comes after deploying the gear means they can do damage in many more ways. The danger doesn't come from allowing Chinese engineers or factory workers to touch a product during its construction. Most electronic products already contain components made in China. That may present risks, but they are manageable. The problem is when someone else can disrupt your flow of parts, monitor your communications, and throttle or terminate service on your network, whether it supports a hospital or a military exercise.

More important than the location of the factory is the question of who calls the shots at the equipment manufacturer. What's the culture of the company? Is it that of a dynamic Silicon Valley innovator? Does it have the paternalistic gentility of a midwestern utility? Or does it encourage the "wolf culture" that Huawei's founder and CEO urges on his company's employees, which he describes as "bloodthirsty, working in packs, and resistant to harsh environments"?

There's an additional concern about the expansion of Huawei: that the company's technology is being used to project China's influence around the world and advance the values of the Chinese Communist Party. As Huawei deploys crime-fighting systems to municipal governments—part of its "Smart Cities" package of solutions—watchdog groups have raised questions about what is happening to the vast amount of information vacuumed up in those systems. Activists and officials in "Smart Cities" from Kazakhstan to Ecuador have complained that information about citizens may be redirected to servers in Shenzhen. Worse, the danger may not lie in what Huawei is doing *to* its customers, but what it's doing *for* them. Strong-arm governments in Africa have been accused of sweeping up popular opposition party leaders by relying on Huawei-based systems and experts to track them and crack the encryption on their devices.

Huawei denies the claims, attributing them to countries defending their own national equipment makers, explaining away allegations that they "open back doors" in their network switches as misunderstandings over software used in the devices. Huawei asserts their independence from the Chinese government, citing their "private ownership," unlike state-owned Chinese companies. But recent research suggests the company may not be independent from the government at all and, since 2017, China's new

National Intelligence Law requires all companies to assist the CCP's intelligence apparatus.

Is China exporting its own brand of authoritarian surveillance and control? China's views on public participation in governance, as exhibited by the Hong Kong crackdown of 2019–2020, raised the specter of America relying on an authoritarian and belligerent rival to deliver crucial support across multiple critical infrastructures.

• • •

The United States ratcheted up the conflict in 2019, pressing allies to eliminate Huawei equipment from their networks and to ban the company from bidding on future business, but in early 2020 it appeared that calls for cooperation would fail. Then the US made an extraordinary move. Just when the diplomatic pressure on allies seemed to falter, the Trump administration blocked chip sales to Huawei and then blocked sales of chip fabrication equipment to companies who *made* chips for Huawei. China recognized this as a potential deathblow to a company considered essential to the future of the country.

After the United States made these aggressive moves in July of 2020, everything seemed to grind to a halt. The ban on chip sales to Huawei appeared to raise the possibility that they would not only lose the ability to sell to the West, but that they could not deliver world-class solutions to their own captive market. It could be the end of the company. The nuclear option had been played, with great effect but not without terrible risk. Sun Tzu warns that an aggressor should always leave his opponent with an escape route, lest he find himself facing a desperate enemy with nothing to lose.

Without Huawei, what alternative had the free world left itself? North American and European telecom companies had been defeated at every turn as the Chinese system worked on a single, coordinated front to beat the disorganized West. China, with more than a billion people and seemingly a thousand ministries, consistently maintained a unified presence, like the 2,008 Fou drummers assembled for the 2008 Beijing Olympics, drumming as one. They used our freewheeling, every-man-for-himself entrepreneurial market mindset against us effectively. For a long time, it worked.

Ironically, it may be precisely those cultural traits that present the best solution to our current bind. The countries of Europe and the Americas have lost the early battles for 5G, but that doesn't mean they can't win the wireless war.

The remnants of the once mighty wireless telecom equipment sector now mainly consist of Finland's Nokia and Sweden's Ericsson, with limited solutions coming from vendors in Korea and Japan. While these are competent companies with legacies of deploying advanced telecom systems, do they have what it takes to deliver world-class 5G networks? Or does the answer lie in developing systems that allow off-the-shelf hardware from any company to be used in the network, systems that let any company's software engineers—not just those from Nokia, Ericsson, or Huawei—write whatever software is necessary for whomever needs it? China's Huawei has dominated because of the benefits that come with scale, whether in R&D or production, but perhaps the era of scale—at least *company* scale—is about to pass. Is this the beginning of a technology revolution in wireless like the internet revolution that replaced huge, centrally developed and managed phone networks with the even larger and vastly more flexible distributed networks of today's internet? Could the might of a $120 billion Chinese manufacturer be surpassed by thousands of small, medium, and large companies working in a completely new kind of market? The scale we need to win may come from an ecosystem of established players and entrepreneurs that is, together, far bigger than even the Goliath Huawei.

• • •

Many people have written books about China's efforts to assert dominance in economic, technological, military, and political spheres. They talk in terms of marathons and stealth warfare, both of which capture essential elements of the Chinese Confucian philosophy, with a dose of Sun Tzu's advice to "win without fighting."

Wireless Wars isn't just a book about the global struggle between China and the rest of the world. It's about how ordinary people—men, women, executives, scientists, salespeople—as well as politicians and spies, made choices, with the result being the evaporation of the world's leading network equipment companies and the emergence of Huawei as the lone telecom superpower.

This book tells the stories of the people who first went to China to open partnerships with local vendors like Huawei, taught them the basics about making and selling the gear, and sometimes stayed long enough to turn out the lights as the Western parent companies collapsed. It shares the personal accounts from the scientists and executives who were tempted by cheap, smart labor and massive markets, only to find their gains short-lived. It describes the telecom sector's clash of China versus the world through the eyes of people who were there as it unfolded and paints the picture of how we got here, what our options are, and what the consequences may be if we get it wrong. Finally, it proposes a path forward that may enable the United States and other free countries to reap the enormous benefits of next-generation networks while keeping our networks secure from the hands of a geopolitical foe.

The stakes are high. Implications and risk go far beyond mobile phones, or even the enablement of advanced services. They are macroeconomic. Geopolitical. Life and death. The disputes won't be settled by the International Telecommunication Union's standards committee, or even the World Trade Organization. If we don't get this right, they may be settled by the Pacific Fleet.

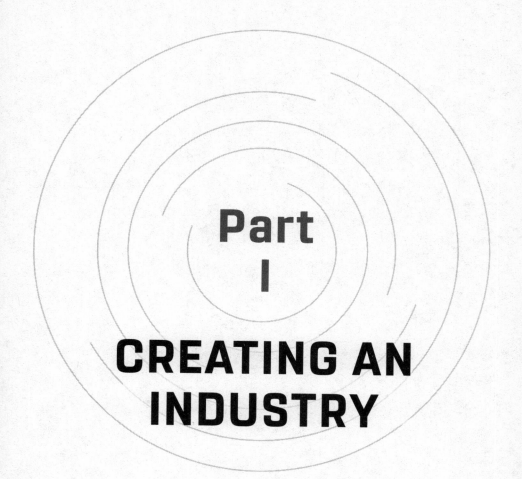

Part
I

CREATING AN
INDUSTRY

The Study

Market potential for cellular communications
appears very limited.
—McKinsey & Company report, 1985

The need for mobile phones is not new. The first car phone dates to around 1901, before the word "radio" existed and years before the first Model T rolled off Henry Ford's assembly line. Although some stories attribute the first known use to a Swedish engineer, Lars M. Ericsson, founder of a telephone manufacturer bearing his name, that's not quite correct; it was his hand-picked successor as CEO, Axel Boström, who made the first call from a car phone.

Boström was interested in cars but struggled with the low reliability of the early prototypes and lack of access to repair services. Finding that he was often stranded on the side of the road and in need of a horse and carriage to get home, he placed one of his company's wooden crank phones in the trunk of his steam-powered car, along with two wires and a stick. When he found himself stranded, he would use the stick to hang the wires over the copper phone lines, crank his phone, and connect to the nearest exchange to complete his call. Certainly not wireless, but mobile, and a car phone by any description.

It's also an illustration of Boström's ingenuity and vision, which would push Ericsson to become an innovative leader in the creation and delivery of communications equipment in the twentieth century. Under Lars Ericsson's charge in the early years, the company's research and development consisted of purchasing telephones from more established manufacturers, such as AT&T Bell Laboratories, then run by Alexander Graham Bell, and reverse engineering them, producing competent, cheaper knockoffs that took advantage of Bell's failure to adequately secure patents in Sweden. When Boström became CEO, he grew the company, exploiting lower-cost labor by moving Ericsson's manufacturing offshore to cheaper locations like Beeston, England. As sales slowed in his own country, he turned to export markets, hiring aggressive overseas agents to take business from his more expensive competitors.

Ericsson may have been the first company to use these tactics to gain ground on better-established foreign rivals, but it wouldn't be the last . . .

THE BIRTH OF WIRELESS

By the middle of the twentieth century, Bell's company had grown into the industry's—and the world's—undisputed technology leader. In the early 1960s, scientists at AT&T conceived of a new way to let people make calls from their cars, far better than that clever approach of Boström. By covering the country with towers, each connected by copper wires to the central switching office, placed in the middle of a honeycomb-shaped cell site and equipped with radio transceivers, they could enable people driving in their cars to make phone calls as if they were on a wired network. Crude versions of "radio telephones" had been around for decades, but they employed a different approach, were difficult to use, had awful sound quality, and could serve only a few customers in a city at a time. Now, these technical breakthroughs promised the quality and ease of a wired call, and even allowed the user to maintain the call as they moved from one tower's coverage area to the next—a "handoff," the scientists called it—which opened the service to support long drives as well as brief calls from the middle of a traffic jam. The technology sprang from inward-looking scientists, not outward-facing marketers, so no evangelist burst forth to tell the world that "our lives have changed!" But they had.

If AT&T realized the impact this invention was going to have, they didn't show it. No unanticipated technical hurdles emerged to throw water on the concept. No business-case economics appeared to make them think this was a bad idea. It wasn't. And yet, little effort went into commercializing the technology over the coming decades. AT&T was a regulated monopoly for nearly a century, earning a profit that was fixed by the government and determined by the size of the company's assets, not by the brilliant inventions they could bring to the public. As a result, the company's efforts went into providing gold-plated service to the American public, as defined by the company's engineers.

Over the next two decades, the world's most brilliant scientists, working at Bell Labs' headquarters in Murray Hill, New Jersey, continued to refine and improve on their model, testing the service in a small trial run out of Chicago. On April 3, 1973, a senior engineer at another Chicago-based company, Motorola's Marty Cooper, called Joel Engel, a rival at AT&T, and told him that he was talking on his new invention, a handheld wireless phone, courtesy of Motorola Labs. Engel may have been surprised, but what should have served as a wake-up call for AT&T passed, and the company went back to sleep. It was another seven years before the company realized that they needed to make a decision about whether to develop this potential wireless market.

Was it time to turn this magic into a commercial reality? Was this the next big thing or just another passing fad? The company had continued to pour millions of dollars into basic research on the electronics and had teams working on how to arrange elements of the network—the architecture—in order to best provide service. Now they had to decide whether to make the investment needed to provide these mobile services to businesses and consumers, or perhaps just manufacture handsets and network equipment that would allow other companies to provide that service. Or maybe they should walk away from this market altogether.

The cost to end customers of a car phone or one of Cooper's brick-sized handheld mobile phones would be high, and the per-minute rate for calling might be, too, but that paled next to the cost AT&T would have to incur to build a nationwide network, replete with cell towers dotting the cities and highways of the country, crammed with stacks of radio transceivers, processors, heating and cooling gear, and backup batteries. Unable to determine the economic potential of their invention, AT&T's senior leaders brought in the experts.

THE STUDY

McKinsey & Company is one of the world's most prestigious consulting firms, widely acknowledged as the place to go when faced with a complex business challenge. Their associates are the cream of the Ivies: bright, hard-working, polished, and able to tackle any problem that can be answered in a PowerPoint deck. McKinsey had long been an integral part of the decision-making process at AT&T, advising on reorganizations and business operations. But their analytical approach was not without its shortcomings, and they faced a daunting challenge: It's hard to build a forecast without any historical data points to look back on. In 1980, although it had been seven years since Motorola had demonstrated the first handheld cellular phone call, it would be another three years before this prototype became a reality and the first handset would be sold commercially. If McKinsey's predictive compass was off by one degree, the forecast for twenty years out would be meaningless—or worse.

That turned out to be the case with this project, which later became known infamously within AT&T as simply "The Study."

This presentation came at a time when the company was debating whether to pursue wireless services. The team of McKinsey analysts had crunched the numbers to determine if AT&T should become a part of this new industry. They looked at the demographics of the addressable market: How many businessmen drove to work? What portion of them earned enough to put a $5,000 phone in a $15,000 car? What were the forecasted annual sales of Mercedes S-Class sedans and Cadillac Fleetwood Broughams? How bad was traffic expected to get on the country's urban streets and suburban highways?

When McKinsey was ready to present its analysis, sitting at the far end of the massive, polished conference room table, wearing his best navy suit, pressed white shirt, and red-striped tie was a junior executive, tall, lanky, dark-haired. Jim Brewington. Brew, as he was called, was the guy who had been running the Chicago trial, and had just taken over the company's tentative exploration of cellular consumer products—mobile handsets, mainly.

The first time he had ever heard of wireless phones was in 1976, when he was working on something called "New Services" (AT&T was clearly an engineering company, not a marketing firm), and among a long list of projects were these early car phones. AT&T was providing the service more as an experiment than as a business, and even then only to a small number of wealthy executives. Connecting a call was a chore, and sound quality was awful.

Over the next few years, Brew had remained involved in the fledgling service before receiving the invitation to come to the company's headquarters in New York for the McKinsey presentation. He was the closest thing the company had to a subject-matter expert, but at the time Brew was about four or five levels too junior to be in the same room with this crowd, let alone speak up. He was expected to sit in the back and keep his mouth shut, which he did. It wasn't easy.

Brew and the suite of executives watched as the McKinsey partner placed each of the seventy-two foils, transparent plastic sheets with graphs and bullet points, on the overhead projector—and sealed AT&T's fate: The future market for mobile phones would be miniscule, they said. Use of the phones would "cap out at 0.5% penetration" of the market, even assuming universal coverage and awareness. Brew remembers them stating that the market wouldn't exceed 976,000 customers by 2000.

The calculations were no doubt done accurately by the McKinsey whiz kids. But the premise was terribly flawed. The original sin lay in defining the market as businessmen willing to pay several dollars a minute to talk to their offices from their car using a phone that cost thousands of dollars. They didn't grasp what would happen when these forty-pound suitcase-sized devices were miniaturized to fit in the palm of your hand, and they failed to develop an understanding of phones as a fundamental tool of communication that would be useful for parents calling their kids, or plumbers connecting with their dispatch offices. They didn't seem to consider that once a network was built, the phone company's marginal cost to connect each call was close to zero, and the price to consumers might someday approach that.

Perhaps the greatest flaw was their reliance on customer surveys and interviews to determine interest in using this transformational, as-yet-unmarketed product. McKinsey's research used "conjoint analysis," a complex and powerful analytics tool that discerns preferences and hidden desires. According to a faded photocopy of their presentation, McKinsey interviewed 3,178 business executives to gauge their likelihood of using a cellular phone, and they questioned 4,533 consumers on their desires. Based on their answers, the consultants developed a forecast for the future acceptance of cellular phones, accurate "within a 90% band." They were confident in the predictive value of their surveys—far too confident, because although people had answered McKinsey's questions, they didn't know what they were talking about. As Henry Ford allegedly remarked, "If I had asked people what they wanted, they would have said 'faster horses.'" People can't give a useful opinion about their interest in a technology they haven't imagined, especially if it is evolving in ways that even the inventors can't predict.

McKinsey reported that the businesspeople and consumers surveyed felt "beepers and payphones are sufficient for [their] needs." The report went further, cautioning against hopes that the service would pick up if it got better and cheaper: "Significant incremental market growth will not result from changes in either price or other product service attributes." Perhaps hardcore business executives, fighting bumper-to-bumper traffic in their Fleetwoods, would bite? No chance: "Even high potential segments of the population demonstrate very limited interest in cellular communications, at any price."

The conclusions seemed beyond question—but perhaps that was the consequence of not knowing what to ask. While AT&T was still investigating the potential business, Motorola, more confident in the future of mobile telephony, had spent the ten years since their first prototype refining and improving the technology, putting more than $100 million into preparing an offer for the market. Shown a copy of McKinsey's deck forty years later, Marty Cooper has an interesting observation. "You know what they never got?" he asks, shaking his head with a wry smile. "Wired communications and wireless are two different businesses. Wires connect one place to another. Wireless connects a *person* to another *person*."

AT&T PASSES ON MOBILE PHONES

McKinsey would produce several reports for AT&T over those critical first few years of the mobile phone industry, each piling on the pessimism and confirming the initial results of a market that had no future. During one presentation, a McKinsey senior partner, silk suit and salt-and-pepper temples, looked across the boardroom table and confidently assured the leadership of the world's largest company, "The methodology and results of these projections have been validated both internally and externally." AT&T didn't need to bother with the car phone niche. The recommendation came in loud and clear: "Shut it down."

Perhaps McKinsey can be forgiven for dismissing the potential of radio communications. It wasn't the first time it had happened; in 1886, Heinrich Hertz first demonstrated the existence of electromagnetic radio waves capable of traveling through the air at the speed of light. He pronounced his discovery "of no use whatsoever."

As it turned out, McKinsey was also off by a bit. By 2000, there wouldn't be nearly a million mobile customers in the world, there would be closer to a billion. Penetration didn't top out at 0.5 percent; it approached 100 percent in many countries, and even exceeded that in some. And this wasn't just for the wealthiest countries. By 2018, more than 70 percent of the people in sub-Saharan Africa would have their own mobile phone. In the boom years ahead, nearly 976,000 new customers would be adding service *every business day*. Penetration by 2020 exceeded the 0.5 percent cap significantly, when the Ovum Group estimated the total number of mobile connections surpassed 118 percent of the world population.

But most people in the AT&T conference room, on that fateful day in 1980, could not have imagined such a future. The executives gathered up their papers and left the room with their minds returning to the work at hand. In the back of the room, as Jim Brewington watched them leave, the wheels started turning. *This is not a business to kill*, he thought. *This is the future*. But he was in no position to stand up and shout, "Stop!" Addressing the company president would be like a lieutenant interrupting a four-star general with his opinion.

It wasn't long before the world—and eventually AT&T—realized the value of the market. Until the company seized its destiny, the exit from the handset business and the services side of mobile telephony left AT&T far behind its global competitors, just when the opportunity was greatest. This ill-conceived decision hampered and delayed the company's later efforts to build the world's biggest and best wireless equipment business, but it didn't stop them. However, as it became clear how wrong they were, it did derail McKinsey's gravy train at the world's largest telecom company.

In the following years, Brew would rise through the ranks of the company, eventually creating and leading their multibillion-dollar wireless equipment operation, and he kept a copy of The Study in his office. As McKinsey tried to regain its reputation and work its way back into AT&T's roster of consultants, they would periodically send a junior partner to share his thoughts on whatever issues were facing the business. Brew would play a little game with him. ("I know it was cruel," he admits, "but . . .") He would be sitting in his chair with his back to the door as the partner introduced himself.

"Hello, Mr. Brewington. My name is . . . McKinsey believes . . . We think there's a new opportunity to . . ."

Brew would pick up his copy of The Study, spin around, and slam it down on the desk in front of them. "You see that study?" he would yell. "That cost us twenty billion fucking dollars! What do you think about that?! You gonna do another one of those for me?!"

His estimate of the cost wasn't far off. The debate had been raging in AT&T's boardroom about whether they should bet on wireless or invest in computers. In the aftermath of the McKinsey study, AT&T went with computers, first developing its own, later augmenting them with acquisitions. Meanwhile, several fledgling wireless operators had been accumulating licenses from the Federal Communications Commission to operate in markets across the United States, and some of the startups had been accumulating even more debt in the process. In passing on wireless, AT&T missed an early opportunity to snap up the struggling McCaw Cellular, a pioneer national wireless network, for a few hundred million dollars, rather than the $12 billion they ended up paying for it in 1994. Instead, Brew suffered as AT&T spent nearly $8 billion to acquire NCR Corporation ("The cash register business, for God's sake!"), which would be unwound and written off shortly thereafter. Between the money wasted on computers and the lost opportunity to enter wireless cheaply, AT&T had a $20 billion hole to climb out of just to get into the market.

No company in the world could have laid a claim to being a more capable steward of this new service. If AT&T—the company that had invented lasers, stereo, fiber optics, the transistor, the communications satellite, and now cellular service—wasn't going to lead, and even dominate, the market for mobile equipment and services, then who in the world would? AT&T's Bell Labs had invented the technology, but their scientists had invented a lot of things they were never able to monetize, like "laser tweezers" that could pick up an individual atom. Their scientists had even discovered the origins of the universe in the Big Bang, but never cashed in on that either. No company played a greater role in advancing America's interests through technology. But, at the time, the United States, like AT&T, let this opportunity slip through its fingers.

LIBERTY AND
BELL LABS

AT&T was more than just a technical jewel for the country. Perhaps nothing better symbolizes the extraordinary role its Bell Labs played in American culture than a little-known story from this time period. It was 1985 and the centennial celebration of the Statue of Liberty was fast approaching. President Reagan had decided to speak at the dedication of the refurbished monument on Independence Day, but the US National Park Service had bad news for him. The corrosion to her surface, resulting from long-ago modifications of the torch, was worse than expected. Large sections of copper were in bad shape, leaking water, and even threatening the integrity of the superstructure.

It wasn't clear if repairs could be made in time for the ceremony that July. In searching for a solution, the Park Service turned to the country's leading expert on maintaining outdoor copper—the phone company. With over 827 million miles of deployed copper running from the sewers of New Orleans to the Pacific Northwest coast, AT&T and its research arm, Bell Labs, knew more than anyone in the world about what made copper corrode and how to preserve it. The labs had an entire division filled with material scientists who did nothing but test copper's performance in difficult weather conditions. Their farm in Chester, New Jersey, even had fields filled with planted phone poles, the wood treated with different creosotes and preservatives to determine which materials worked best and lasted longest at keeping that copper in the air.

AT&T made John Franey, an expert from Bell Labs, available to the Park Service to tackle the problem with the statue's cladding. Franey spent the better part of a year puzzling over how to repair Lady Liberty's robes and found that, while it wasn't hard to patch the corroded sections of the surface with new copper, the biggest challenge was laying on sheets that wouldn't *look* new. If the damaged sections were patched with shiny copper, they would stand out like a new penny against the familiar green patina of the statue.

Franey first attacked the problem like any scientist would: What chemical process could he apply to new copper to make it resemble the green, weathered material? He was able to develop a solution that appeared to age the copper and create a patina that would match the green of the statue. But he felt uneasy about his clever chemical trick. This wasn't just any copper. If the solution didn't work as expected, he couldn't just generate a work order to roll a truck out to some roadside phone pole. He would be responsible for making Miss Liberty look like she was wearing a patched, shabby robe. And he didn't want to take the chance that the treatment would look good now but corrode too fast, or that the patina wouldn't last.

The answer to Franey's problem didn't end up coming from his lab—at least not from *inside* it. Driving into his Murray Hill office one day, Franey happened to look up at the roof of the Bell Labs headquarters. For more than fifty years, the facility, where Dr. William Shockley had invented the transistor and engineers had developed the sonar systems the US Navy used to defeat Nazi subs, sat under a massive copper-clad roof. And that copper was the same thickness and quality as the copper used on the Statue of Liberty, faded to a green patina that was indistinguishable from her own skin. As Franey drove up to the parking lot, workers were prying the copper sheets up to make repairs on the rotting wood sheathing beneath. Franey smiled—he had found his solution.

A few weeks later, Bell Labs had stripped the copper from the roof of its Murray Hill headquarters and loaded it into trucks to be delivered to Liberty Island. Bell Labs had, once again, come to the service of the United States of America, literally wrapping itself around the Statue of Liberty and preserving the beacon to hold her torch over the New York Harbor for the next century.

Thirty years later, bulldozers took down Building 1 at Murray Hill, demolishing Bell Labs' empty and idle research space—a victim of aggressive foreign competition, market shifts, and bad management decisions. This time, any valuable metal in the roof was sold for scrap. At least Bell Labs' two-million-square-foot Eero Saarinen–designed facility in Holmdel, New Jersey, the lab that brought the world fiber optics, had a better fate when the company downsized and vacated the building. A relatively successful example of America's shift to a service economy, that facility was converted from a center of groundbreaking telecom research and development to a mixed residential, business, and retail complex also available for wedding receptions.

AT&T Creates a Business

> Always remember that it's much easier to
> apologize than to get permission.
> —Grace Hopper, American computer
> scientist and US Navy admiral

Although AT&T exited the mobile phone business—the handsets—and stayed out of the service provider business as well, they maintained a small unit that made the gear that service providers placed into cell towers to connect wireless calls to the network. The management team knew they would need a stubborn, energetic person to run this wireless equipment division, especially given the lack of commitment in the executive suites.

DECOY OFFICE

Shortly after the McKinsey presentation, Jim Brewington had been transferred to Denver to serve as the western sales vice president for AT&T. He liked living out west, which reminded him of his upbringing in Idaho. He had led a classic farm boy's life: Since he was fourteen years old, he had helped out on his family farm in the mornings and evenings and worked as a foreman on a nearby ranch during the day when he wasn't in school. Like all farmers, Brew learned to be resourceful, and he understood that sometimes you just need to get the job done. Those traits would serve him well.

In his sales role, he quickly established himself as the company's top seller of equipment for the fledgling cellular business. He also built a reputation as a renegade, rarely seeking permission and ignoring convention and company culture when they got in the way of achieving his business objectives. "When I was made a supervisor," Brew says, "I learned that you could take the power—all you had to do was *take* it. If you want to be successful, you gotta go do things. You can't wait for things to happen. *You gotta just go do it.*"

The cellular systems Brew was selling were cobbled together from other product lines and were technically unimpressive. But the market was beginning to make traction. Even with the limited investment—or perhaps because of it—AT&T was losing $100 million a year, and the division's group executive, Bill Marx, decided he needed someone who could turn things around. Marx was a good boss for Jim: no-nonsense, willing to break the rules, and—most importantly—willing to run interference for a member of his team who was similarly inclined and able to deliver the goods.

It was a Thursday when Marx called Denver.

"Brew, we gotta do something about this wireless business," he said. "I'd like you to come back here and run it."

Brew was not eager. "You know, Bill, I got the customers liking me. One of them even takes me to the Super Bowl. You got a lot of other hotshot guys, why don't you give one of them a chance?"

"Take some time and think about it," Marx replied.

Six hours later, Marx called again. Same conversation. So Marx said, "Brew, give it some thought. Why don't you sleep on it?"

The next morning, Brew was just getting to his desk with a cup of coffee when the phone rang. It was Marx, of course.

"Brew? Effective Monday, you're being made an officer of AT&T. You're gonna head up wireless for Network Systems. We need someone to straighten it out, and that someone is you. The chairman will be looking forward to meeting you at 8 AM Monday to hear your plan. Don't be late."

Brew thought about his plan on the flight to New Jersey that Sunday. One thing was clear: In a huge, bureaucratic company like AT&T, he would not be able to turn the division around if he was constantly asking permission to make changes. When he arrived, he sat down with Marx and told him what he needed.

"If you really want me to make this into something," Brew said, "I got to pull it all together as a unit. I can't do this unless Bell Labs reports to me, the factory reports to me, product management reports to me." This was heresy at the company, especially the idea that Bell Labs might allow its scientists to report to a business unit head. It just wasn't done that way.

Marx didn't hesitate.

"Done," he said. "But you are going to have to talk to Ian Ross." Ross was the president of Bell Labs. "They're going to swallow when they hear this."

Brew worked his changes through the bureaucracy and reorganized all the components into a single unit that reported to him and over which he had complete control. Next, he overhauled the factory, turning its Columbus, Ohio, facility into a world-class manufacturing plant.

The headquarters for this new division would be in Whippany, New Jersey, a temporary location hastily built during World War II to house military research for the War Department. There, Bell Labs scientists had conducted work on projects like inventing guidance systems for the supersonic Nike missiles and developing towed-array sonar used to track Soviet submarines. The building had already outlived its expected life, and the floors and thin walls rattled as people walked through the halls, but the ample space allowed the wireless unit to work under one roof.

It was a far cry from the plush suites at the executive offices in Morristown, New Jersey, where Brew's office was located. As a newly appointed officer of the company, he was entitled to the perks that came with the job, and this location was one of them. But that building was huge, and there was half a mile of hallways between his office and his boss's suite. So Brew had an idea. He outfitted his "official" office, maintaining the appearance of an occupied workspace, and rigged his desk phone to forward calls to Whippany. There, he secretly set up his *real* office, where he could get his hands dirty building the wireless product line.

In a matter of months, he had the operation humming, free from the interference of corporate bureaucrats and absolved of the need to pass each decision through half a dozen business unit presidents.

And then, one morning, he got a call.

"Brew, Bill Marx. I need to talk to you about something. Can you come down to my office?"

"Sure," Brew replied. He calculated how long it would take to get down to his car, drive to HQ, and find his way to Marx's office. "Give me about half an hour—I'm just cleaning up some work I'm buried under."

"Really?" said Marx. "Well, I'm standing in your office right now and your desk looks remarkably free of any kind of work, as well as any executive."

The gig was up, and Brew had to let his boss know what he had been doing. Fortunately, Marx understood his desire to focus on the product teams, not sit through administrative meetings. And Brew was making progress, which is what mattered. Marx let him know he would have his back.

It was becoming clear, especially in overseas markets like Hong Kong and Europe, that this mobile thing was no niche. Growth was explosive, and people were using the mobile phones in ways that were never anticipated. While AT&T had nearly missed the boat on mobile service provision, it looked like their wireless division might yet become a player on the equipment side.

3

Retaking the Lead

Creating the future is more challenging than playing catch
up, in that you have to create your own road map.
—Gary Hamel, American consultant

By 1996 AT&T had a new problem. Their Communications Services division was selling wired and wireless connectivity to companies and consumers, and their Network Systems division was selling the equipment that AT&T and its competitors used to deliver those same services. It became clear that AT&T's equipment group would have a tough time selling equipment to competing service providers as long as AT&T's services division was in the same business. As a result, AT&T split itself into several independent companies, including a service provider that kept the AT&T name, and a set of equipment makers, one of which was renamed Lucent Technologies. Lucent retained AT&T's network equipment products, including the fledgling wireless business that was run by Jim Brewington, and took with it most of Bell Labs.

While AT&T, now focused entirely on communications services, was able to leapfrog competitors in the wireless services business with the belated acquisition of McCaw Cellular, Lucent didn't have it as easy. This new operation sold network equipment, which went into cell towers and operating centers, to carriers like Verizon and former parent AT&T. Lucent had missed years of development work around cellular services and couldn't simply acquire the needed technology from another company; there were no start-ups or small equipment makers to snap up.

Lucent had to do something to grab the attention of their wireless customers, something to show off their latest technology. One idea was to introduce a new version of mobile telephony that had technical advantages over both the GSM (Global System for Mobile Communications) standard being deployed around the world and the multiple standards being deployed in the US. But catching up with the European vendors wouldn't be easy.

CHASING EUROPE

Europe mandated the GSM standard by law—no other mobile technology platforms could be deployed on the continent—and as a result GSM enjoyed great economies of scale there and throughout the world. GSM used a technique called "time division" to cram more calls into the available radio spectrum, which created some quality problems but allowed the use of inexpensive processors. Along with the scale economies, this made for a very cheap handset and cell site. But cost was just one reason why mobile phones took off so quickly in Europe.

When cellular services were first introduced around the world, they were appreciated as a convenience for consumers in the United States. But they were a godsend for people in Europe, where it might take a month or longer to get a wired line installed at your home or workplace. Europeans signed up for mobile phones as an alternative to lousy, expensive, unreliable wired service, hard-to-use pay phones, or no service at all. As a result, wireless took off much faster in Europe than in America, and the quality of coverage quickly outstripped that in the United States.

This fed the coffers of the European equipment makers, transforming Alcatel, Nokia, and Ericsson, as well as some sleepy European "national favorites" like the UK's Marconi and Germany's Siemens, into serious technology leaders. These companies began to build up an outsized presence around the world, where wireless took off even faster in less developed places with worse wired services, and the Europeans even gained a foothold in the United States. For Americans, it was hard to imagine anyone outside their borders providing telecom equipment and services better than their own companies, but when it came to deploying wireless services, Europe left the rest of the world in the dust. By 2002, when Italy was passing 100 percent teledensity—the number of phones per person in a country—for mobile services, the US was barely at 50 percent. It was no wonder that, for most of the networks in the world, the European standard was embraced and deployed.

This ascendance of European telecom vendors didn't sit well with Jim Brewington and his band of developers in Lucent's Product Realization team. It was bad enough to see local, well-regarded competitors like Motorola and Nortel growing their business with the US carriers at Lucent's expense, but watching America's regional Bell companies turn to Nordic countries—*Finland and Sweden?*—for their high-tech solutions was too much for this Idaho boy. Brew had a plan to put his company back into the game.

SWEATING ON THE TARMAC

A famous photograph shows Marty Cooper making the first handheld cellular call to his rival at Bell Labs on April 3, 1973. The phone in his hand was a prototype DynaTAC, and news coverage of this breakthrough was extensive, as excitement grew about the transformative technology. But it took more than ten years before Gordon Gekko would lug around the first commercial version of that phone, dubbed the "brick," in the 1987 movie *Wall Street*. This was a sign of how promises would outstrip reality in that sector, right up through 5G.

Few market segments have been as overhyped as the mobile communications market. People want to believe that they can put a device in their pocket that will transform their lives, and suppliers are all too happy to overstate, exaggerate, and outright misrepresent the services that are available or just around the corner.

Keenly aware of this aspect of the industry, Brew made a decision to jump ahead of the European competitors by hyping his own wireless technology.

Qualcomm, a California-based chip company, had invented a new standard called CDMA (code-division multiple access), which used "code division" to pack the calls in. While it delivered superior quality and capacity, it did so by requiring more advanced processors, which cost more to build. It wasn't tried and true like the GSM standard, but it provided an opportunity for Lucent to leverage a leading American chip maker, along with innovation at Bell Labs, to bring a better solution to market—better clarity, more capacity, and (best of all) the chance to grow margins by charging more for a superior solution.

Nothing makes a technologist more nervous than a live product demo. But for Brew's team it was practically the only option. "We knew we were behind," says Dave Poticny, Brew's head of engineering. "We had lined up Verizon as a first customer, but we were wary of any missteps under so much scrutiny." They needed to get some good press and show off their breakthrough. So as Lucent readied its first CDMA product, Poticny briefed Brew on his plan.

"Brew, we have to demo this, not just to the public but to the media, the analysts, everybody who influences opinion."

Brew thought about it. After all, this was a new product launching into a rapidly growing, cutthroat market.

"We need to be careful," Brewington said. "Look at Motorola's demo in Hong Kong—it was a disaster. We have one chance to get this right!"

Poticny agreed and came up with the idea of holding a big press conference where he would present each of the reporters with a shiny new cell phone—the latest Motorola StarTAC flip phone—which they could test on Lucent's new network as they learned about the company's plans for the technology. But first, Poticny needed to find the right place to set up the test network. Every country in the world had its own rules on who can use what part of the radio spectrum—you can't just turn on your base stations in another country to try them out unless you can find a local carrier with a license for the spectrum to cooperate—and that left few options. His team got busy scouting locations, keeping in mind the top criteria that an industry reporter would seek in covering the launch of a breakthrough technology solution: nice beaches and a world-class bar. Preferably with a golf course. The team eventually landed on Puerto Rico.

Poticny explains, "We built two sites at the furthest east end of the island, covering the grounds of the Conquistador hotel. And we had two sites in San Juan, covering the airport and a tourist part of the city, where the reporters would get to visit a famous castle." His team knew that coverage would be fine in these locations. But the thirty-mile bus ride between the hotel and the city?

"That was a problem," Poticny says. "We didn't have coverage along the highway that would take them to the city—we just didn't have time to build it out." That meant that, during the bus ride, the phones would stop working. Poticny pictured the bus pulling out from the resort and watching the smiles fading from the reporters' faces as the cell tower's signal faded. *Can you hear me now?!*

No, that couldn't happen. Poticny needed to make absolutely sure the reporters would not turn their phones on between locations. Not an easy challenge, given the cool phones they would be playing with. But Poticny had an idea.

On the day of the product demo, the reporters were delighted to be presented with the new toys. As expected, the tech experts marveled at the ease of calling and the sound quality as they dialed their girlfriends, husbands, colleagues to show off. Then it came time to make the trip to the city. Poticny guided the group to a side door off the hotel lobby and looked out to the parking lot.

There was no bus.

But there *was* a sleek executive helicopter.

"Let's go," Poticny said. Smiles all around, not least of all on Poticny's face.

The group filtered out into the heat and climbed aboard the helicopter. And then, Poticny lied to them. He told them something that was unsupported by any empirical evidence, undefendable by any of the thousands of researchers working in Bell Labs' wireless facilities. But it was a lie that would be repeated millions of times more in the coming years, to the annoyance of road warriors and business travelers the world over: "Please turn off your phones before we take off. It's not safe to operate them while in the air."

The group of professional skeptics and tech experts looked up at him in surprise and concern. Suddenly these cool new toys felt important and dangerous in their hands. They meekly turned them off and closed their phones, careful to avoid causing an air catastrophe.

The rest of the weekend went perfectly. With a bit of scrambling, Lucent had launched its efforts to catch up with the mobile industry Bell Labs had invented. The gambit—and the CDMA bet—paid off.

"Once we got the lead, we never let go," Brew recalls.

At least, not until the wheels came off the whole company.

Losing the Plot

You can have a certain arrogance, and I think that's fine,
but what you should never lose is the respect for others.
—Steffi Graf, German tennis player

While Brewington was getting Lucent's wireless equipment business rolling, his former parent company was starting to staff some of its own growth businesses. AT&T chairman Bob Allen needed someone to take over a small group and help him figure out if a new market—the internet—was worth getting into. So in 1996, Allen brought back Dan Hesse, an executive who had been running AT&T's equipment business, labs, and factories in Europe, the Middle East, and Africa (EMEA). Hesse was considered a rising star in the company. A graduate of Notre Dame, Cornell, and MIT, he possessed what *USA Today* described as "Jimmy Stewart–like earnestness." He was six-foot-five, with an avuncular demeanor that would later grace Sprint's TV commercials when he ran that company. But Hesse was no kindly caretaker—he was intensely competitive and determined to grow the company's new businesses quickly.

GETTING LAZY

Following his time on the special project (AT&T grudgingly decided this internet thing might be real), Hesse was made CEO of another nascent but rapidly growing business, AT&T Wireless Services. After a career that had him selling telecom gear to the European carriers and leading a successful pitch to win a massive Saudi contract, he now found himself a customer of the equipment he used to make. The AT&T Wireless unit he took over didn't use equipment from the former AT&T Network Systems, now operating as Lucent Technologies. Lucent had been spun off from AT&T, so Hesse had the freedom to buy network gear and handsets from any of Lucent's competitors, including US-based players like Motorola or Nortel, or those in Europe, like Nokia, Alcatel, or Ericsson. (No companies from Asia had yet made the consideration list, even as second-tier suppliers.) Before Hesse took over, the unit was an "Ericsson shop" because of the legacy equipment contracts that came with McCaw Cellular.

Hesse was not surprised at founder Craig McCaw's choices. In his eyes, Ericsson had emerged as the global leader, with Nokia close behind. He was not impressed when he considered the options of bringing business back to the US vendors. While the European companies used GSM, US equipment makers had mostly settled on a variation called TDMA (time-division multiple access), which was just different enough from GSM to lock out competition from European equipment makers. CDMA, while impressive, was just getting started with Verizon and Sprint, and was considered a risky option.

After the North American vendors had trailblazed the analog 1G systems, "2G/GSM network gear was a new frontier," Hesse says, "and the US vendors had lost leadership to the European vendors. The American vendors were competing using proprietary standards like TDMA, which was a tweaked GSM, and CDMA as a way to set themselves apart, not necessarily as superior offers."

He found the US vendors were also falling well behind the rest of the world in the handset market. Motorola, which had created the first mobile handset and then stormed the market with their world-conquering StarTAC flip phone, was coasting on past successes but failing to recognize that the market was moving quickly to digital phones with multiband radios that could work on different systems. Hesse learned this the hard way when he spoke to Motorola's product heads and requested a multiband StarTAC that would support a planned new offering from AT&T.

"My customers don't need a digital device," Motorola's product lead told him.

Hesse couldn't believe this response from Motorola and recalls his shock: "The arrogance was striking. What do you mean, *your* customers? Those are *my* customers! If I decide not to carry your phones anymore, your sales go to zero!"

That year, 1998, Hesse launched a new calling plan—the Digital One Rate—that changed the way mobile phones were used by eliminating charges for roaming and long distance. This was a game changer at the time: customers could now use their phone wherever they traveled in the country and not worry about the high roaming and long-distance charges that were typical. But for it to work, users needed multimode phones that could switch between the traditional, analog networks and the new digital ones. The old StarTAC couldn't do this.

Motorola's sales to AT&T customers dropped off a cliff.

Motorola CEO Chris Galvin jumped on the company jet to see Hesse. Galvin met Hesse in his Kirkland, Washington, office and said, "Dan, my guys screwed up. They didn't listen to you. Can you let us back in?"

Hesse pulled out his new Nokia 6160 and waved it at Galvin. "As soon as you can make one of these, we'll talk."

Motorola did eventually develop a multimode phone, but the fall from grace hurt the company, and their cellular equipment business never regained its place. While their handset business was able to rebound and have another period of success, they significantly exited the market for network gear a few years later.

One American maker down, two to go.

LISTENING IN

Dan Hesse's request to Motorola was not an idle one. There were a lot of reasons why he wanted to move past the first-generation analog base stations and phones to digital systems: capacity, cost, features. But there was another concern with the technology, one that hadn't made much news yet but was about to.

The early 1G analog networks left calls vulnerable to casual eavesdropping. Literally anybody with a radio scanner could listen in on anybody else's call if they were near the same cell site. And many of the American carriers were slow to upgrade to the more advanced 2G digital systems that automatically encrypted calls between phone and tower.

This tardiness was exploited by a married couple in Florida in a crime that, for the first time, opened eyes around the world to the security risks inherent in mobile telephony. In December of 1996, John and Alice Martin recorded a conference call between Newt Gingrich, then Speaker of the US House of Representatives, and his team of Republican strategists, as they worked through a response to an ethics investigation. John, a school janitor, and Alice, a teacher's aide, used an off-the-shelf police scanner to listen to the conference call, picking up the signal from then chairman of the House Republican Conference and a future Speaker himself, John Boehner, who was on a Florida vacation. Unfortunately, this member of the Republican team, sitting in his car in a restaurant parking lot, was late to the 2G party and hadn't upgraded his handset. The Martins just happened to have a portable recorder handy for capturing messages to their not-yet-born grandchild, or so they claimed, and they held it up to the radio's speaker, recording the whole sordid conversation and turning the tapes over to Democratic leadership in the House, causing a Republican crisis to get worse.

The Martins were fined $1,000 for violating federal wiretapping laws and for turning over the illegally obtained recording, and the event would serve as the first public example of the threat, not just to one man's reputation but to national security, that could come from compromising a cellular network.

If a couple of folks in Florida could procure damaging information on someone who was considered the second most powerful man in the US government, using gear available from Radio Shack, what could a motivated spy agency do?

ACHIEVING SCALE

The protections that US telecom vendors had put in place through proprietary standards were starting to limit the companies they were designed to protect. Hesse saw it as a matter of scale: "The US didn't have the economics that came with 2G/GSM, which everyone knew would become the global standard." Those economies of scale meant that component makers could price for hundreds of millions of handsets, not tens of millions. Base stations would be made by the hundreds of thousands, not tens of thousands. Prices would drop accordingly. Likewise, the scale of a single global standard meant that more features would be offered, and there would be more handset variations and more options on the infrastructure side. (Years later, as the entire world converged on a single 4G/LTE standard, the benefits of scale would become even more pronounced.)

Although the United States had created the network gear market, a series of bad choices was causing American equipment makers to fall behind, and they feared that the Europeans might seize the initiative and run the US manufacturers off the tracks. As a result, vendors like Lucent and Nortel redoubled their efforts to beat Ericsson and Nokia, the perceived global leaders.

It's said that sometimes a slow-moving train can hide a fast-moving train coming up beside it. That was the case here, and the fast-moving train would be China—but not for a few more years.

Poverty and the Power of Communications

It doesn't matter whether the cat is black
or white, so long as it catches mice.
—Deng Xiaoping, former leader of the
People's Republic of China

Around the same time McKinsey & Company was giving their presentation in AT&T's lushly furnished conference room, another consultant
named Colin Golder found himself in Beijing standing in front of a cold,
bare hall filled with engineers wearing identical blue Mao suits. It was 1980,
and he was attempting to explain how to build a digital telecom network in
China, a country that was decades behind the rest of the world in deploying
communications networks.

STARTING FROM SCRATCH

Golder grew up south of London, where he had completed secondary school
and taken a job as an apprentice technician at the organization that would

become British Telecom. While there, he enrolled part time in a technical training institute. But his employer recognized his talents and sponsored him for a university scholarship that eventually helped him earn a Master of Science in Telecommunications Systems.

Still, Golder wasn't the kind of consultant that McKinsey would hire. He wasn't used to climbing out of a Lincoln Town Car and looking around to find the nearest Four Seasons or Ritz-Carlton. He worked in the trenches. Literally. He had spent more than a decade engineering switching systems and consulting for countries that were burying the copper wire that would form their first telecom network.

His work took him across the Middle East, sometimes on the payroll of the host country's telecom ministry and sometimes working for one of the United Nations' development organizations. Most recently he had spent two years working for the deputy minister of telecoms for Saudi Arabia, helping Philips, the Dutch conglomerate, and Ericsson to create a telecom network in a country that had almost none. When Golder's work for the Saudi ministry wound down in late 1979, he took a role with Philips. Almost immediately after settling into his new office in Hilversum, the Netherlands, Golder's boss walked into his office with a question.

"You've worked for the United Nations, haven't you?" he asked.

"More than once," Golder replied.

"Well, we just accepted an assignment under the UN's Funds in Trust program and we need someone to go to China." The program allowed deep-pocketed benefactors to fund specific directives that were administered by the UN. In this case, China had approached the UN and asked to have a number of telephone companies send representatives to deliver lectures on the state of the art for wired communications systems. In other words, it was looking for a face-saving way to learn how to build a digital telecom network and find out what the major equipment vendors had to offer.

"How does Beijing sound?" Golder's boss continued. "Just for a month. Give them lectures, tell them how to build a phone network. Teach them about the kind of things you were doing for the Saudis."

Golder hadn't spent time in China, but looking back, he remembers how the idea appealed to him. Mao Zedong had died a few years prior, and Deng

Xiaoping had taken the reins and began an extraordinary process of opening up the country.

A month later, Golder arrived at Beijing's brand-new international airport. Terminal 1 had just been completed, expanding capacity from the tiny original airport terminal, which had been reassigned to handle private jet travel for VIPs. It was underwhelming. This glorious new airport terminal for the world's most populous nation was not much bigger than a food court.

It had no food court.

Golder was met at baggage claim by a Communist Party member who would be his host for the coming five weeks and was escorted past the lines in the airport, through a diplomatic channel, to a waiting government car. It was already late and the drive to Beijing was dark, but made more so by a strange behavior Golder witnessed in the driver.

"He kept turning his lights off," Golder recalls. "Then he'd turn them on for a while, then off again." Was he signaling someone along the way, letting them know a foreign guest was in the car?

Golder finally asked his host, and the answer he received shed light on the challenge he would face educating the country on digital communications technology. His host explained that an edict had gone out from the Party to "save battery power." It was determined that people could help achieve this by turning their car lights on and off when driving at night.

When they arrived in Beijing, Golder was met by a city that was starting to recover from the economic devastation caused by thirty years of Mao's People's Revolution. Maybe not starting *yet* but realizing that it needed to start. Beijing had no Ritz or Four Seasons, but Golder's host took him to the Minzu Hotel on Xichang'an Street, almost dead center in the city, which would be his home for the next month.

It wasn't much to look at, but he was impressed with the quality of its construction. The Minzu had been built in 1959 as part of a command project, ordered by Mao himself, in tribute to the founding of the People's Republic of China. Ten magnificent structures were commissioned to showcase China's glorious successes in the Great Leap Forward. Though a mere 161 feet, the ten-story Minzu hotel towered over all of them and stood as the tallest structure in Beijing at the time. Its name—*min zu* means "minority national"—claimed a noble purpose, to host the fifty-six indigenous ethnic groups of China. In reality, the hotel had been built for someone exactly like Golder to use while doing precisely what he was there to do.

And over the coming years, many of China's minority ethnic groups would find themselves hosted by the government in far less luxurious locations.

LIFTING A NATION

Golder checked in and went to his room. In the morning, from the large windows, he looked out over the smoggy, gray city. As far as he could see sprawled the *hutong*, the narrow alleys and the low-slung brick buildings that seemed to be about the only structures in the city, serving as home, shop, restaurant, and business.

There was a single, narrow car lane, more than adequate for the few vehicles on the road, all driven by government officials. The rest of the wide streets were jammed with chaotic bike and foot traffic. In 1980, Chinese people still aspired to *sanshengyixiang*, "three rounds and a sound": a wristwatch, a bicycle, a sewing machine, and a radio. Humble aspirations, but still out of reach for most. All organized industry had been wiped out by Mao's Gang of Four—the small cabal responsible for the worst excesses of the Cultural Revolution—and it showed.

Golder went downstairs to wander around the neighborhood near the hotel, and he found himself dodging bicycles piled seven- or eight-feet high with products being delivered throughout the city. At least in Beijing, that one round, the bike, seemed to be present in numbers. And though there was hardly a car to be seen, the air was thick with the smell of soot from the half-burned coal that ran Beijing's power plants, the same coal that was used in most homes to cook pork or to boil water. The dust was everywhere, and the air stank of it.

Golder decided to explore a bit, roaming through the *hutong* and getting an up-close look at the ancient city. The first thing he noticed was that he was practically the only Westerner on the streets. The crowds seemed to be made up entirely of Beijing's local citizens, dressed in drab, loose-fitting Mao suits. As he walked by one of the low-slung buildings that doubled as a restaurant and a family's home, he couldn't resist reaching out to touch the wall. Feeling the eyes of the locals upon him—the *laowai*, the "old foreigner"—as they stared and then turned to each other in amusement, he traced the edge of a brick.

"Even the bricks seem to be handmade," he recalls thinking, "not quite square, like they were made in the backyard."

When Golder returned to his room at the Minzu, he walked to the large window and looked out again over the chaotic city, letting out a whistle. "These guys could use some help!"

There was no question about it. Deng had committed to improving the quality of life in China, but he had a long way to go. In 1980, China had a population of just under one billion people, of which more than 80 percent lived in extreme poverty. China's GDP per capita stood at $195. The problem wasn't that it lagged behind the United States, at $12,575 per person. The problem was that it lagged behind Sudan, at $392 per person, and all but a handful of the poorest countries in the world.

It would be essential for China to build a national telecom network as the first step in industrializing the country and raising the national standard of living. That network would be the first crude brick in the foundation of Deng's new China. The link between more telephones and improved prosperity was backed up by numerous studies. Increasing teledensity caused the income of the people in the country to go up, with an improved standard of living, reduced child mortality, higher education, extended average life span, and all the other markers of a wealthier society.

"Deng seemed to have worked that out himself," Golder says. "I suppose it's why the UN sent me there."

COFFEE MAKERS AND BUTTONS

The morning after his arrival, and each morning after that, Golder was picked up at the hotel by one of the few cars in the city, all of which seemed otherwise reserved for leaders in the Communist Party, and taken two miles to the telecommunications building next to the Forbidden City.

Golder remembers the first time he walked into the room and looked out at his audience. "There must have been fifty engineers sitting there—including many women—wearing nearly identical Mao suits," he says. "Short hair, serious faces." Not a lot of joking going on.

Golder spent five or six hours a day standing in front of the group, talking about the state of the art of modern telecom networks and answering their questions, breaking only for a midday trip back to the hotel for lunch. As for his personal time away from the meetings?

"I'm sure I was being watched every minute of the day."

One evening Golder stopped by a room at the Minzu that doubled as the Philips consumer division's "Beijing branch office." He knocked on the door and saw the light coming through the peephole darken. There was a pause. Then the door opened wide to show a fellow European—Johannes, another employee of the Dutch conglomerate who was in Beijing pursuing a different business purpose. Golder had just introduced himself when he sensed a movement behind his colleague. A young Chinese man was sitting at a small table, staring at him. Golder looked at him, then back to his colleague, then back to the man at the table. Finally, he leaned in closer and asked in a hushed voice, "Who's that?"

"Oh. That's my assistant," Johannes replied. "Provided by the Chinese government to help me out in any way he can. He's here all the time. *All the time.* Come on in."

Though initially puzzled by the assistant's presence, Golder eventually figured out that China had a deep suspicion of all foreigners conducting business in the country and made sure to assign an "assistant" to each one to keep an eye on all comings and goings. Golder never visited the room without finding the assistant there, and Johannes assured him that even when the gentleman left for the evening, he still didn't quite feel like he was completely alone.

Johannes was starting small, doing his best to sell refrigerators to Chinese buyers, while also trying to get Chinese manufacturers to make the glass jugs for a Philips coffee maker. Simple glass carafes, heat tempered, with a handle. It wasn't going well. At the time, the Chinese manufacturers had no capability to build even a basic product that met European standards for consistency, price, and quality.

Golder thought the Chinese market seemed pointless. Certainly, China wasn't capable of manufacturing any technology that could be exported, with its well-deserved reputation for poor quality and careless design. And while the country had the need to consume whatever goods the West might export, it wasn't clear that it had the buying power to make the market worth pursuing.

"I remember when I first heard them explain what the bait was for Western companies," Golder recalls. Early on, one of the administrators from the Trade Ministry had made the pitch to Johannes as they sat in a quiet teahouse near the hotel.

"You can see that the people here do not buy clothing like they do in Paris," said the administrator. "Yes, we are a poor country. Why would we be worth talking to? Perhaps each person owns one suit"—he gestured to the gray Mao suit that was still the nearly ubiquitous uniform at the time—"and maybe he needs to buy three shirts a year."

The administrator was hardly making his case.

"But each of those shirts has seven buttons," he continued. "That means each person here in China must buy three shirts with seven buttons every year. That's twenty-one buttons a year for a billion people. Would a button company be interested in selling twenty-one billion buttons a year?"

Johannes got it, and so did Golder. But the country that bought twenty-one billion buttons a year also shared one phone line for every five hundred citizens. It had a long way to go.

6

The Price of Admission

The Contract Is Signed. And Now the Negotiation Begins.
—*New York Times* blog headline, on doing business with China

It wasn't until the 1990s when developing countries—making up 75 percent of the world population—began to acquire significant communications systems. As they did, it became clear that wireless phones would serve the purpose better than fixed lines.

The math was compelling. As expensive as a cell tower might be, and as much as an early mobile handset cost, it was still cheaper than digging a ditch from a local phone switch to a person's home or business, perhaps a mile away, and running a few thousand pounds of copper cable to connect the two. For many countries that lagged on teledensity, the wireless explosion completely bypassed wire-line service. More than 4 billion people would eventually transition from no phone service to mobile handsets without even owning a wired phone.

And among all those developing countries, China led the pack in rolling out service. From its position in the 1970s as a country with fewer than one hundred thousand phone lines serving one billion people, it started adding lines at breakneck speed. Including the traditional wired service, China began adding more lines per year than the rest of the world combined.

This meant that even for companies struggling to fulfill surging demand for mobile service in developed countries like the United States, the Chinese market could not be ignored, and China knew this. In keeping with its long-term plans to upgrade its own ability to compete, China began to place strict requirements on anyone hoping to sell into the country. Among the top rules: foreign companies could not bid for more than 20 percent of any in-China build-out, and all bidders needed to establish in-country joint ventures that would include technology transfer and process sharing. China intended to use the eagerness of foreign companies as a catalyst to create a domestic Chinese telecom industry that could serve their own need for equipment and maybe, just maybe, turn around and sell something to countries elsewhere in the world.

The Western suppliers were all too happy to participate in this game.

FROM THE MIDWEST TO THE FAR EAST

For one young manager, the game would become more of an adventure. David Heard, a twenty-three-year-old graduate of Ohio State, had just been moved over to AT&T's Columbus, Ohio, location. He was thrilled to work at the world's biggest—and best—maker of communications equipment. Cellular communications was the hottest place to be, with sixteen million phones already in use worldwide, and explosive growth that would turn that number into over seven hundred million by the end of the decade. Heard had just been put in charge of a new project—moving production of AT&T's cellular base stations from North Carolina to the Ohio factory, which was better suited to ramp up to meet the surging demand of the wireless service providers. He was happy to land a job with a top company, even if it meant working in a regional factory instead of one of the corporate offices AT&T had across the country. Heard knew that if he wanted to have a successful career at the company, it wouldn't hurt to earn his stripes working on the shop floor.

On the day that would mark the beginning of his adventure, in the fall of 1991, the Columbus factory looked like an industrial version of a flea market, with massive machines stacked at odd angles and forklifts squeezing crates between aisles. Heard felt a hand on his shoulder—a large one—and turned to look up at Bill Robinson, his boss. Robinson stood at six foot five and wore Coke-bottle glasses that looked about right for someone who had graduated at the top of his class from MIT.

"You got a minute?" he asked.

"Sure," said Heard.

"You're going to China."

Heard squinted. The thought of a trip to China appealed to him. Among other things, it meant his boss thought enough of him to send him on a business trip overseas. "Okay. When? And for how long? I'll need to get a passport."

"No. *You're going to China.* To live there. Look, this wireless shit is exploding. We're nowhere in Asia, in the China market. We're going to set up a partnership with a local Chinese telecom equipment factory. Go negotiate this joint venture. You'll be the manufacturing guy to say yea or nay about what we do."

"Cool," said Heard, feeling anything but. "Uh. I don't speak Chinese."

"Don't worry about that," said Robinson. "We have two guys from Ohio State lined up to come by your house every night after work and teach you Mandarin. They're good. Six weeks should be plenty of time to learn. And don't worry about housing. We'll set you up in a room at the China World Hotel. Nice place. Brand-new. In fact, I think everything nice in that city is brand-new." Robinson smiled and turned to walk back to his office. "I'll be down at Donerick's Pub tonight if you want to swing by and talk about it. Can I assume your answer is yes?"

Heard had just finished earning his MBA by going to classes in Dayton full time during the day and then working the night shift at another local AT&T factory. The idea of now spending his evenings studying Chinese didn't bother him. He already had a reputation for unbounded energy.

The offer was a no-brainer. "If I wanted to run this company someday," he says, "I needed international exposure. It's Career Planning 101." He was ready for the challenge, but he would quickly find out that his MBA hadn't prepared him for the experience he was stepping into.

A couple days later he connected with Jim Brewington, Robinson's boss. Heard knew that Brew was a legend at the company, having kept them alive in the wireless business after "The Study" from McKinsey sought to pull the plug.

Brew, meanwhile, liked Heard, who had an approach to business that was pragmatic, like his own. And he was dogged. Just get the job done. He expected Heard to become one of his stars, and he thought the China assignment would give him a chance to shine. In fact, there weren't a lot of people he could trust with this kind of a project. Too many of the people in the organization had "Bell-shaped heads," as he called it; they thought like Bell System lifers, the product of a hundred years of monopoly operations, delivering gold-plated service as defined by their own labs. And he needed someone with the right personality. Confident verging on cocky. This job would need energy, smarts, and a confidence that wasn't necessarily supported by any empirical evidence.

Brew briefed Heard on the assignment. "Just give them the analog shit," he said. "The stuff we're forklifting out of Ameritech." The industry had reached its first transition, where the original first-generation gear, the 1G systems that the wireless service trailblazers like Ameritech had deployed, was ready to be replaced by 2G digital equipment. The services wouldn't be terribly different for customers, but the capacity was much greater for the network providers who were already seeing service quality hurt by network congestion. "Don't even think about letting them have the digital gear." Those digital base stations were for the American providers—high technology that AT&T was hoping to put into US networks.

Factories were at capacity, and Brew liked the idea of taking the old analog gear that was being scrapped, pulling it from the customer's cell towers across America, and sending it to China, where a local factory could refurbish it and put it into networks in Guangzhou or Beijing. Securing and negotiating that arrangement would be Heard's job.

The plan had three benefits, in Brew's view. One, it opened access to a potentially massive market for AT&T. Two, it allowed AT&T to unload gear that would otherwise be scrapped. And three, it avoided any advanced technology from being delivered into the hands of a country that Brew distrusted and feared would steal the company's intellectual property. Brew was deeply suspicious of the Chinese government and its intentions toward Bell Labs' inventions. His instincts were right: years later, in 2017, the bipartisan Commission on the Theft of American Intellectual Property would estimate that the annual cost of IP theft to the US exceeded $225 billion and "could be as high as $600 billion," and that China "remains the world's principal IP infringer."

Brew wanted to make sure Heard understood the dangers. He leaned forward. "Now, once you get there, you're going to have to watch what you say and where you say it. Especially with your phone calls back here. We'll send you over with one of these." Brew pointed to a box on his desk that looked like an oversized answering machine. There was no internet to worry about; all files were held locally and wouldn't be sent back and forth overseas. But he assumed that any phone call made from the hotel-based offices would be monitored.

The device he pointed to was a scrambler—just a commercial-grade device, not something the NSA would use for top-secret communications—but Bell Labs had designed it, and the odds were that no one on the Chinese side could descramble it, listening equipment or not. "Fire this up when you want to talk with us about anything but the weather," Brew said. "It takes a few minutes for the two devices to shake hands and sort the encryption/decryption out. And then we can assume we're secure. You still want to do this?"

Heard smiled. "Let's go."

FOR QUALITY
PURPOSES

AT&T had learned of China's commitment to surveillance during their efforts to sell 5ESS switches, the company's bread-and-butter phone switch that sat at the heart of every voice network. They were deploying switches to a Chinese client right on schedule until one day when the orders dried up. They didn't just slow down; the purchase orders for these million-dollar switches went to zero overnight. AT&T had seen orders dip before, when there was a quality problem, or the buyer was trying to sweat them for lower prices. But this time the Chinese buyer offered no complaints and had no requests for the baffled AT&T sales team.

Then, in an internal meeting discussing the puzzle, one of the product guys suddenly spoke up. "Hey, wait a minute," he said. "I think I may have an idea about what's going on here. A few years ago, one of our guys working in the country, Colin Golder, told me about a tour he took at the Guangzhou phone company, back when they were using an old ITT 7E rotary switch, the one they started making in the fifties."

The product guy recounted how Golder had been shown the room where the switch was located and found a clerk hunched over intently, listening through a pair of headphones and scribbling notes on a pad. The clerk had seemed pretty settled in, like this was a full-time job. Looking up, he started when he saw the Western visitor, his eyes widening as he slowly laid down his pencil. Golder's host had swept him away from the door and moved him back into the hall.

"For network quality purposes only!" he had explained. The company was merely listening in to the calls to make sure the lines were clean and clear.

The product guy explained where he was going with the story: "We're backlogged on monitoring gear." This was the equipment that carriers used when they received a court order for a wiretap, known as "lawful intercept" gear. "If you look back over the purchase history, for every voice switch that was sold the Chinese team had bought monitoring gear to track callers." In other words, China would not install a single switch if they couldn't listen to what was being said over it.

This discovery showed a trait about the Chinese authorities that would later cast a pall over the lives of everyone living within China's borders, as well as create panic among carriers in the Western world. But, at the time, this was great news! The team back at AT&T's Oklahoma City factory cranked up production of the lawful intercept gear and immediately the sales of the million-dollar switches returned to plan.

It didn't take long after arriving for Heard to settle into the pattern. Negotiations began at 7 AM sharp. They continued through the day with a short lunch break, another for dinner, before winding down at 11:30 PM. This repeated seven days a week.

Occasionally they would talk about the cultural differences between the countries. "We hear you give your factory workers eight days off every month," one of the Chinese hosts said.

"Yes," Heard replied. "We call them 'weekends.'"

But mostly the talk was all business, and it went on for months.

If the local team thought they could grind Heard down, they had the wrong man. This was someone who, later in his career with AT&T (and after the spin-off, Lucent), would book trips to Europe that didn't include hotel rooms. He would fly in; shower and change in the airport lounge; spend the day in meetings, the evening taking the clients out to dinner, all night drinking with his local team; have a car drop everyone off at their apartments; then drive straight to the airport for his flight to the next city. He kept a tub of peanut butter in his carry-on so he could grab a spoon, flop into his seat, and get enough protein, fat, and calories to tide him over. Then, eyeshades on, seat back and asleep before takeoff, getting just enough rest to propel him through the next day to do it again. Brewington had picked the right man for the job.

The biggest problem Heard faced was the constant pressure from the Chinese team to work with AT&T's newer digital gear. His partners didn't like the way China was being treated, and the nature of the "joint venture" was insulting to the hosts, who felt that a Chinese factory could be doing a lot more than refurbishing used last-generation cast-offs from an American cellular carrier. More than once they would harangue him to open up more of the company's technology and move some R&D to the Chinese mainland. Their interpreter would struggle with the metaphors: "If you want to be our long-term partner in China, we have to be like two grasshoppers connected by a string. Neither can escape from the other, neither is superior."

The pressure for more advanced technology increased, and after weeks of negotiations it began to look like the mission might be a bust. His Chinese partners insisted that the deal include the new digital systems coming out of Bell Labs. Heard wouldn't yield. He knew that China needed to get its own wireless networks deployed and operational, and if it couldn't get its hands on AT&T's new stuff—yet—at least it would learn about the business processes of the world's leading communications technology firm. It would have to do.

Meanwhile, the Chinese telecommunications ministry was keeping tight controls on what deployments it allowed the Americans to bid on. While AT&T limited China to refurbishing used analog equipment in their joint factory, the Chinese government in turn limited AT&T to bidding on no more than 20 percent of the nation's build-out. The rest of the business, whenever at all possible, would be provided by local suppliers. The joint factories helped get around this somewhat by creating a win for both sides, and the fact was that even 20 percent of this market was enormous. Years later, one of AT&T's early customers, China Mobile, would manage a network with more than 940 million cellular subscribers. This was not small potatoes.

The talks continued. Heard tried to explain that the services enabled by the analog gear were the same delivered by digital radios. Call quality was similar; features were nearly identical. The early 2G network providers played up supposed differences, but the end user would hardly notice. To be fair, digital signals reduced power consumption, meaning the batteries (and thus the phones) could be smaller and the talk time longer. And the reduced need for bandwidth meant that carriers could manage capacity better, which translated into fewer blocked calls and better sound quality.

Heard spent the next eight months in the grind of daily negotiations, using his crash course in Chinese mostly to greet people at the start of the day, comment on the lunch as best he could, and wish his hosts a good night come closing—his language efforts sometimes drawing smiles that Heard couldn't always identify as polite or derisive. He and the legal team relied heavily on their translator to interpret each side's positions throughout the day, making an already long project that much more tedious.

The AT&T lawyer who sat at Heard's side through the process was a Texan who looked and sounded the part, with a deep Texas drawl. For the first several months, he and Heard sat in the conference room at the China World Hotel, slugging it out with their Chinese hosts as they negotiated over intellectual property, pricing, ownership of the facilities, and lesser details. Whenever AT&T's interpreter was out of the room, the negotiations would pause, and the lawyer would chat with Heard or return to his paperwork. Unlike Heard, he made no effort at pleasantries in the local language, giving his hosts the impression that he didn't think enough of them to even try.

One day, late in the negotiations, as the teams were closing in on final numbers, the lawyer spoke up. "Can we take a time-out?"

"Time-out?" asked Heard. "Yeah, sure."

The interpreter translated the "time out" as best she could to the Chinese team, and the discussions paused while the lawyer and Heard stepped into a side room.

"Time-out? What does that mean? And why now? We're starting to get somewhere."

"Exactly," the lawyer said slowly in his deep voice. "Here's what you need to do. They need this many points for pricing, and this many points for sales volume . . ." He proceeded to explain to Heard how the opposing team was evaluating their objectives and what they needed to hear in order to accept a deal.

Heard stared at him. *"How the hell do you know that?"*

"Well," he drawled, "I speak Mandarin."

"What?!"

"Yeah, I married a Chinese girl about ten years ago."

Heard was stunned. "Eight months in that room and not one *ni hao*?"

The lawyer had bided his time and let the opposing team come to the conclusion that these two *lao dongxi*—silly old fools—didn't speak a useful word of Chinese and certainly couldn't understand it. When the interpreter would leave the room, the Chinese team had become increasingly lax and impatient, and they had begun discussing their intimate details at the table. Perhaps these Americans weren't as dim as the Chinese partners thought.

Less than a year after Heard landed in China, the deal closed and the factory opened.

The Birth of Huawei

Since the Opium Wars, China has been defeated
by countries with less population, wealth, and
geographical resources again and again. One of
the root causes . . . is our inferior technology.
—Xi Jinping, General Secretary, Chinese Communist Party

Foreigners were not the only ones who recognized the value of China's massive telecom market. When the Chinese government began a push to develop an in-country telecom industry, Ren Zhengfei smelled an opportunity. He had a strong technical background, having served as a midlevel army officer in the People's Liberation Army's engineering corps, but he had only limited business experience, working for a few years after his discharge for a state-owned oil company. In 1987, he left the oil gig, raised $5,000 in seed money from a handful of friends and associates, and launched the company that would become the world's largest manufacturer of telecom equipment and smartphones.

A DIFFERENT PATH

Huawei's first activities were humble, distributing fire alarms from a Hong Kong–based manufacturer, and reselling Private Branch Exchange (PBX) switches—which were used to connect office desk phones to each other—in rural Chinese markets. Yet just a few years after founding the company, Ren decided to forgo the joint ventures with Western firms that nearly all the local companies were lining up.

"A lot of the joint-venture models worked against the Chinese companies," says Eric Harwit, a professor of Asian Studies at the University of Hawaii and an expert on China's economic advance and the spread of communications technologies in the East Asian region. "The Chinese partner would become dependent on the Western partner and not innovate themselves." Ren sensed, correctly, that these foreign firms would be wary about sharing their best technology with the Chinese upstarts, leaving a short horizon for any Chinese firm that built a business around aging technology controlled by non-Chinese companies. So instead of depending on outsiders to drive his company's technology road map, Ren made a fateful decision: Huawei would develop its own switch.

According to an executive familiar with Huawei's initial business in Hong Kong, the first "original" product Huawei made was a version of the Mitel PBX they had been reselling from Hong Kong. The source, familiar with both companies, says that Mitel decided it was easier to join Huawei than to fight them, and they began supplying Huawei with the components needed to make the PBX work well.

From there, Huawei invested heavily in R&D, staffing up and developing, by 1993, the first large switch completely made by a Chinese company. Its capacity was in the thousands of phone lines, not close to a world-class switch, and it lacked the more sophisticated software—advanced billing capabilities, new features, and so forth—that was expected in a commercial-grade phone switch. But if it was a Huawei-designed switch, the technology still came from what the company described as "reverse engineering" of Western switches rather than a true innovation of its own. Still, these simple Huawei switches served as adequate solutions for a market that was, in places, still using hand-cranked phones.

This investment very quickly paid off, establishing Huawei as the leader among domestic manufacturers and securing a contract to build out the first nationwide telephone network for Ren's old colleagues at the PLA. China's military leaders felt there would be grave risk to allow a foreign company to install switches in their network and have its eyes on communications vital to the national security of the country. Instead, they chose to go with a local, trusted vendor and suffer whatever expenses and growing pains were needed to ensure more secure communications.

Shortly after signing the government contract, Ren met with Jiang Zemin, general secretary of the Chinese Communist Party and president of the People's Republic of China, and had an exchange with him that summarized the urgency of securing the country's communications from foreign threats. Ren claimed in an interview that he told the Chinese president "that switching equipment technology was related to international security, and that a nation that did not have its own switching equipment was like one that lacked its own military." He said that Jiang agreed with this assessment, replying, "Well said."

BOMBS AND THE BELL SYSTEM

Ren and Jiang aren't wrong: When two countries descend into all-out armed conflict, one of the first things each tries to do is destroy the other's ability to communicate, to take out the enemy's "command-and-control" abilities. Around the time Huawei was first setting up the company, a chance meeting on the other side of the world showed the significance of telecom switches to a country's national security, especially in wartime.

On May 23, 1991, as the graduating class of the New Jersey Institute of Technology waited in their seats to be awarded their degrees, a recent graduate from the class of 1986, Lt. Robert Wetzel, stepped up to receive the Trustees Achievement Award. Next to him on the podium was Bell Labs scientist Robert Lucky, who was receiving an honorary degree to add to his PhD in electrical engineering from Purdue. They were two Jersey boys named Bob, but they discovered that they had a lot more than that in common.

A few months earlier, on January 17, 1991, Lt. Wetzel's Grumman A-6 Intruder had taken off from the deck of the USS *Saratoga* headed for Iraq and tasked with dropping its MK 20 Rockeye cluster bombs on "target H-3." As the first Gulf War began, the US was determined to cripple Iraq's ability to communicate, coordinate defenses, and manage the response to the coalition forces' efforts to drive them out of Kuwait. One of the key ways to do this was to destroy the nation's communications infrastructure. The Pentagon set out to bomb every phone switch, radio station, and TV studio in the country.

As Lt. Wetzel was still a few miles out from his target, a surface-to-air missile took out his starboard engine. Wetzel and his bombardier/navigator pulled their ejection handles at two hundred feet, racing over the desert floor at nearly five hundred miles an hour. Knocked unconscious, Wetzel parachuted to the ground for a hard landing and was held among the first prisoners of the Gulf War.

His colleagues had better luck. Over the coming days, they launched thousands of bombing sorties to turn Iraq's command-and-control facilities into smoldering rubble.

But many of these missions didn't need to happen. A few months before the start of the war, a group of men appeared at Bell Labs with a request. The Pentagon wanted to disable the Iraqi national communications network, shutting down their command-and-control capabilities during the war that everyone knew was coming. Bob Lucky, one of the most respected "big thinkers" at a company that was filled with them, was assigned to the project. He assembled a team composed of people who had made their careers figuring out how to keep networks running but, in this case, they would do their best to figure out how to destroy one. Not long after getting the assignment, Lucky reached back out to the DoD.

He had a solution.

If the Pentagon could get password access to the Iraqi network, the Bell Labs team had developed a series of commands that could disable the phone system completely. Better yet, at the end of the war another series of commands would restore service.

The Pentagon considered this elegant solution, but ultimately passed and went with an approach they were more familiar with. They launched thousands of bombing sorties, destroying the communications infrastructure with what is delicately referred to as "kinetic action."

Iraq's ability to react and respond to the invasion was significantly compromised by the destruction of its telecom network, which was no doubt instrumental in the coalition forces winning the war. As the US would learn two decades later, a software solution that allows an occupation government to restore the system with a few keystrokes could be an invaluable tool when it comes to winning the peace.

There's no doubt that the planners at the Pentagon and at the National Security Council made note of the idea that a relatively modern telephone network effectively could be destroyed though a software hack.

Thirty years later, Dr. Lucky won't confirm his role in this project for the Pentagon. "I have no idea if it's true or not . . ." he says, however, "I remember doing some exercise about disabling the Iraqi phone systems and I suggested flying over the country and randomly dropping backhoes, which would inevitably be used by farmers to cut the fiber optic lines . . ."

With assured revenue from the large government contract, Ren was able to fund a push into rural regions of China not served by foreign companies or Chinese manufacturers. And it's here that we begin to see the birth of Huawei's aggressive measures to line up customers. According to Professor Harwit, Huawei would set up partnerships with the managers of local, government-owned phone companies and pay "dividends" to the individuals based on the volume of gear they approved for purchase from Huawei. This activity was not explicitly against the law, although it was controversial for obvious reasons. Because the personal interests of the bureaucrats making the local purchasing decisions were closely aligned with Huawei's corporate interests, the company's sales took off, and the company grew to become the dominant homegrown telecom equipment maker. In later years, as the company grew, Huawei took over the local operations and made honest men of these bureaucrats by converting them to managers of Huawei's local branches.

TRAINING ON THE TOWN

While Huawei used unconventional means to win business, the measures seem sophisticated compared to the business development approach reportedly taken by their then rival, state-owned telecom equipment maker Great Dragon.

"Great Dragon had a center where they did training for the carriers' operations people," says Mike Tessler, founder of BroadSoft, a communications software firm that became an industry leader before it was bought by Cisco for nearly $2 billion in 2018. In the late 1990s, Tessler had visited Great Dragon's training center on a trip to meet with his partners on their own turf. "They would bring these guys in from the hinterlands and they would be entertained for the week."

From what Tessler saw during his stay at his hotel, which Great Dragon also used for their other business guests, there wasn't much training.

"These guys would get up at noon, have lunch, then go to bars and clubs all night long," Tessler recalls. "I asked my hosts at Great Dragon, 'Guys, what is this?' The answer was very matter of fact. 'None of them will go back and tell their bosses that they didn't learn anything. And this is how we keep our professional services revenue so high. If they have a problem with the equipment, they need to call us.'"

Tessler didn't understand how Great Dragon got away with it. Didn't their customers' bosses wonder what they were up to all week?

He saw the answer as the visiting trainees were being loaded on the buses home. Each man was handed a box of manuals and spec sheets—their course work for the prior week. They could bring it to the office, show it to the boss, and stick in on a shelf. The most important part of the manual was the service center's phone number on the binder.

Great Dragon—the state-owned national champion, the equipment company founded, funded, and promoted by the Chinese Communist Party to build the country's telecom networks—wasn't interested in learning how it was done in the West. They weren't worried about building a better, cheaper switch. They didn't need to, with the protection of the government. But their products never worked well, and they weren't able to invent or steal anything that did.

Great Dragon dissolved into obscurity over the following years.

TRACKING HUAWEI'S OWNERSHIP

Huawei continued its rapid growth, and as they did, their sources of capital to fund this growth, and the ownership of the company, remained mysterious. Short of finding the actual ownership documents and reading them in the original Mandarin, it would be impossible to penetrate the smoke screen around Huawei's governance and control. Few Western journalists have laid hands on the ten slim, blue-bound volumes listing the owners—all Huawei employees—to review the "official" records. But two academics were able to take a deep dive into the records and laws involving ownership and to present a compelling argument that the company is not privately held at all, at least not in any practical sense.

Professor Chris Balding spent years teaching at the Peking University HSBC Business School, located in Shenzhen. A self-proclaimed Libertarian, Balding prides himself on defending academic freedom and once called out Cambridge University for its willingness to censor articles in its *China Quarterly* journal that were deemed offensive by the Chinese government. You wouldn't think China would be open-minded enough to allow a firebrand like Balding to teach at one of the most prestigious universities in the country.

It wasn't. "There is a profound sense of relief to be leaving safely," Balding told one publication shortly after his contract was abruptly terminated and he was forced to exit the country. "There are many cases which resulted in significantly more problems . . ."

Some time before his unexpected departure, Balding studied the true ownership of his adopted country's largest technology firm. Corporate ownership can be a complicated matter anywhere, let alone in China. Private companies in many parts of the world are not obligated to publish audited documents that spell out the details of their capital table, showing who owns what. Public companies, at least in the United States, are under far more scrutiny and must provide transparency regarding their equity ownership and debt holders. In Europe, Ericsson, the closely held giant in Sweden, has complexity in its own ownership. While publicly traded and owned by hundreds of thousands of individual and financial shareholders, it has historically been *controlled* by two—a Swedish bank and the Wallenberg family. At one point, they controlled over 80 percent of the voting power despite contributing less than 1 percent of the capital, leaving other shareholders with little say in corporate decision-making. Complex, perhaps, undemocratic, sure, but transparent.

In reviewing Huawei's ownership, Balding sought legal expertise from Professor Donald Clarke, the David Weaver Research Professor of Law at George Washington University Law School, in Washington, DC, and a highly regarded expert on Chinese corporate governance. Clarke's fluency in Mandarin helps him cut through third-party assessments of Chinese law, and he has published extensively on the issues that have arisen from China's economic reforms.

Balding and Clarke found that Huawei is 100 percent owned by a holding company of which founder Ren Zhengfei owns 1 percent. The other 99 percent is owned by what is described as a "trade union committee." Balding and Clarke were able to determine nothing of the governance of this committee or its makeup; even the members of what would typically be described as the company's board of directors were not named. As Balding and Clarke pointed out in their analysis, holding companies typically have few employees. So the "trade union committee" that presides over Huawei's shares is not composed of the *holding company's* union employees. As a strictly legal construct, it likely has few employees, or none; the workers—the union employees—would be in the *operating company.*

Balding and Clarke came to this conclusion: Employees do not own actual stock in the company; they hold virtual stock that allows them to share in the profits. They cannot vote and do not control the company or even influence control of it, and even their limited rights are canceled when they leave the company.

Balding and Clarke describe the employee-ownership claim as a masquerade, with "volumes of names and numbers displayed to journalists—paper records, under glass, in a shrine-like setting, at a high-tech company in the twenty-first century—bearing every mark of being a Potemkin shareholder register."

The two professors undertook a meticulous analysis of what this unusual "trade union committee" really means under the system there. Union leaders are not selected by the members, nor are they accountable to them. They are appointed by their company's management—an outrageous idea for any Western company—or by the administratively superior union. Trade union officers in China are accountable not *down* to their members but *up* to the next superior entity. The difference is not nuanced; above the individual company level, union leader salaries are paid by the state. This continues all the way up to the All-China Federation of Trade Unions (ACFTU), which may really have the last word in Huawei's governance. And, according to the study, who controls the ACFTU?

The Chinese Communist Party.

As Balding and Clarke's analysis concludes, to the extent the company is controlled by a union committee, it is controlled by the state—the CCP. And for specific business decisions, Ren has absolute veto authority, which the articles governing the company say is inheritable, for example to Ren's daughter, an odd rule for a company supposedly owned by the employees.

Huawei disputes Balding and Clarke's findings, saying they are "based on unreliable sources and speculations, without an understanding of all the facts." They describe the conclusions as "completely unsubstantiated" and assert that "no government agency or outside organization holds shares in Huawei or has any control over Huawei." Clarke has responded that Huawei has not challenged him on specific facts.

But if their findings are correct, the CCP would have an interest in driving needed capital into the company. They would also have a means: In 1997, as part of a restructuring, state-owned and influenced Shenzhen banks began lending money to employees to fund purchases of "virtual stock" in their employer. This money, which Balding estimates may have been billions of dollars, was presumably funneled into Huawei's operations. It's possible that the CCP established a pathway through the Chinese banking system to ensure that Huawei received the investment capital needed to undertake explosive growth.

And what of the five initial investors—Ren's friends and associates—who put up the $5,000 in seed money to create what would become a multibillion-dollar company? Wealthy beyond their dreams? Angry, bitter, and litigious? Their names aren't known. No record remains.

Part II

GIVING IT AWAY

8

Rising Star

Huawei runs a very disciplined company.
Every desk has a sleeping bag.
—Bob Holder, former AT&T and Lucent executive

Huawei's first real effort at selling telecom gear overseas was prompted by an ill-advised attempt by China to sell one of the few technologies it *was* successfully exporting. Although China publicly took a strong stance against Saddam Hussein during Iraq's 1990 invasion of Kuwait, within a year of the war's end, intelligence agencies learned that Iraqi military officials were making trips to Beijing to procure new anti-ship missiles that could be used against the US and allied navies. These ship-to-ship missiles were not the best and latest from the Chinese arsenal; they were believed to be converted first-generation YJ-6 air-to-ship missiles, first developed by China in the seventies. In effect, they were the equivalent of the 1G analog base stations that AT&T allowed China to use in its own war on teledensity. They weren't likely to take out a nuclear aircraft carrier, but still, the Chinese government appeared to be exporting weapons for the use of the Iraqis, and this created tensions with the Saudis, who felt threatened by the re-arming of Iraq. Both sides were eager to repair the rift.

Enter Huawei, which was just getting its feet under itself and starting to establish business operations as a domestic supplier to the local phone companies across China, especially in the underserved remote provinces. Saudi Arabia might provide an opportunity for Huawei to get its first crack at a true international bid, although with a highly unusual business development approach.

OIL FOR PHONES

In late 1998, Saudi Arabia's crown prince and de facto ruler Abdullah bin Abdulaziz was undertaking an unprecedented tour of the world, traveling to the United States and Europe before visiting several cities across Asia. Crown Prince Abdullah was the highest-ranking member of the Saudi royal family to ever visit China, and his huge entourage made this visit the capstone of the world tour, spending four days in the country meeting with party officials and business leaders. On his meeting with Premier Zhu Rongji, the prince said, "I had an impression of truth and openness during my talks with the Chinese premier." One thing that *was* true was China's growing need for oil, and Saudi Arabia was eager to start selling it to them.

Less compelling was China's value as a supplier of goods to Saudi Arabia. At the time, China was limited to making components that were incorporated into finished electronic and consumer products by industrialized countries, but the Saudi market had little need for anything but finished goods. Nonetheless, in 1998 the two delegations signed a joint memorandum to pursue new avenues for trade between the countries.

China's leaders pressed the prince to commit to specific trade enhancement, and he obliged, at least as indicated in his public statements calling for increased economic cooperation between the countries. It appeared he also placed a word with one of the leaders of his own Ministry of Communications—as Lucent Technologies would soon find out.

GENERALS ON THE WALLS

Lucent was in the midst of a massive Saudi project, deploying $5 billion worth of fixed, mobile, voice, and data networks across the country, and the contract was up for renewal. As with all such contracts, vendors were required to hire a local agent, whose role was crucial—they navigated the arcane Saudi rules and bureaucracy and helped get past whatever roadblocks the foreign vendor faced. The agent also acted as an interface to the "project sponsor," a government official assigned to oversee major projects and often a member of the royal family.

In this case, guidance, perhaps from the crown prince or his entourage, had reached Lucent's agent, suggesting to him that it would be a good start to this Saudi-Chinese relationship if an invitation were extended to a Chinese vendor in general, and Huawei in particular. Lucent may not have cared to bring another vendor into their huge Saudi client, but keeping the customer happy was important, and China certainly presented no competitive threat to the world's greatest equipment maker. The team found a perfect messenger in Colin Golder, now a board member of the Lucent switching joint ventures in Taiwan and China. His long experience in Saudi Arabia was complemented by his knowledge of China, so the team asked him to make a visit to meet with Huawei.

Golder was at his hotel in Taiwan, recovering from a bout of food poisoning, when he received the request.

"I checked my passport and found I had one remaining Chinese visa," Golder recalls. So he called the Lucent team in Beijing and asked them to set up the meeting with Huawei, giving few details about the purpose in advance. Golder dragged himself to the airport, slumped into his seat, and flew into Hong Kong. There, he swung by the local Lucent office to meet with the lawyers and pick up a printout of a boilerplate joint cooperation agreement before taking a hotel car to Huawei's headquarters in the special economic zone in downtown Shenzhen.

As he walked down the halls of the Huawei offices, Golder was struck by the images he saw. Where Western companies covered their walls with pictures of the latest products, or images of founders, leaders, and heroes of the company, Huawei's corridors were lined with pictures of generals and other military officers visiting the building, decked out in full dress regalia.

"It was pretty clear who was funding the place," he says.

When Golder walked into a small room and saw the man waiting for him, he knew that this meeting would not happen—at least not that day. Maybe it was because Golder's business card gave his title as "Head of International Business Development," or maybe it was the world-renowned Bell Labs logo on it, but Huawei had sent their "Head of Development"—the man who oversaw technical research—to meet with him.

It was not a match.

"Here's the story, Mr. Li," Golder explained. "I'm here to explore a joint opportunity. Crown Prince Abdullah has just met with your leadership, and it's been suggested to us that we may want to work with you. Since you're the head of development, I fully imagine you are not the person to deal with this. I suggest I get back in the car, return to my hotel, and come back here to the same room at precisely the same time tomorrow, if that works with Huawei."

Li agreed, and Golder returned the next morning to find a very different representative. He entered the room and was greeted by a woman this time. She was young and dressed impeccably in a Western-style business outfit. She had an air of seriousness and confidence about her.

"Mr. Golder?" she said. "I'm Madam Sun."

Golder recognized her immediately as the newly appointed chairwoman of Huawei. Sun Yafang spoke hesitant English, but with the assistance of an interpreter, Golder explained his intent to help this fledgling company bid for a small piece of the Saudi contract. It wouldn't be easy.

"They were green as grass," Golder says.

At the time, Huawei was just a start-up telecom equipment maker with no global presence. Nearly all its business consisted of selling low-end phone switches to the rural markets in China, servicing local telephone companies that even the other Chinese vendors wouldn't approach, and even those sales benefited from the "dividends" paid to the purchasing managers. If the Saudi bid was successful, it would be Huawei's first win outside the region and mark the beginning of a joint partnership between the world's largest, most sophisticated telecom equipment maker and the upstart from China, ready to finally cut its teeth on a global bid, perhaps opening a path to future international success.

As they laid out the logistics for developing the joint program, Sun made an odd request. "As we work through this agreement," she directed, "please don't use email."

That would be easy enough. Without an electronic version of the agreement they would be marking up a paper document anyway. He wondered why she was averse to engaging electronically—after all, they were preparing a bid for phone switches, not plotting a coup—but he appreciated that it would enhance security, if such things mattered so much to her.

BLOWN COVER

Golder noticed something about Madam Sun over the three or four days they worked together, something he found curious. "Her English improved markedly," he says with a grin. "She must have been working night and day to improve it . . ."

It's possible that she went from speaking almost no English to developing a conversational ability in just days. But years later a document released by the CIA-based Open Source Center describes Ms. Sun Yafang as an operative with the Ministry of State Security, China's primary intelligence agency.

Huawei denies the claim, attributing it to CIA misinformation. That might have been the end of the issue, but unfortunately for

Huawei, Sun's college's alumni association, apparently unaware of her cover story as an executive at Xinfei television factory, raved about her achievements in their April 2006 newsletter. The chatty update, since stripped from the internet but viewable on the Wayback Machine, congratulated the notable alumna on her career, joining the Ministry of State Security after college before leaving to take a role at Huawei.

The Open Source Center document describes Huawei founder Ren Zhengfei's gratitude that, during the company's financial troubles in the mid-nineties, Sun used her relations with the government to secure financial assistance, with significant bailout aid (reportedly in the hundreds of millions then billions of dollars) being directed to the company. As the alum newsletter celebrates, Sun (and the government) "saved Huawei" and established her "extraordinary position" within the company.

The newsletter also explains why her English seemed to "improve" so markedly during her meetings with Golder. Around the time of Golder's meeting with her, Ren had presented his leadership team with a surprise quiz. They returned from Spring Festival to find a TOEFL, the Test of English as a Foreign Language, waiting for them. Ren put his managers to an essay contest, writing about how breaking the rules can help with governance of the company. The winner, writing in perfect English, was Sun, whose essay was entitled "Don't turn the tide of heroes," and encouraged learning more by working closely with competitors, who could be a great source of insight to Huawei.

She meant it.

Over the following days, Golder was impressed with Sun's focus and intensity. By the end of the week, Huawei had its first joint agreement with a major Western manufacturer.

The companies set up a work group to prepare the bid on their piece of the Saudi project, and Golder handed the matter off to Lucent's bids and proposals team in the Netherlands, who provided Huawei with assistance in assembling and submitting its first international bid. Together, they presented their proposal to the Saudi telecom authority and waited anxiously for the response.

The proposal was declined.

In their first effort to make an international telecom sale, Huawei had failed. Nonetheless, they benefited enormously from the process. The Huawei executives got to see how Lucent read and answered an international specification. They learned how to assemble a quote. They learned to not be so . . . literal in their description of current capabilities and compliance with requirements. All companies take some liberties in responding to requests for proposals (RFPs) with their description of existing technology and services. The challenge is to not overshoot your headlights by promising something that won't be ready by the time it's actually needed.

In short, the experience served them well. Madam Sun could have taken some solace if she had known that, while her company lost this small bid to supply Saudi Arabia as a subcontractor to Lucent, over her next twenty years as chairwoman, she would lead the company to more than $300 billion in such wins, including billions in 5G mobile networks.

9

Boom to Bust

Never interrupt your enemy when he is making a mistake.
—Napoleon Bonaparte, French military leader

While Huawei was getting its feet wet, the rest of the 1990s marked an unbroken series of growth years and record profits for the Western telecom makers. Whether selling into greenfield markets like China or adding capacity and features in advanced markets, it seemed the run would never end. But it did, and more abruptly and spectacularly than perhaps any sector collapse since the Dutch tulip bulb crash of 1637.

NOT A BANK

Trouble may have been just around the corner, but in the late 1990s, growth only seemed to be accelerating, as large service providers expanded their networks and a crop of new competitive service providers like WorldCom and Global Crossing raised billions of dollars from investors to create their own networks from scratch. By the end of the decade, equipment makers like Lucent were adding internal capacity—factory space to meet the production demands and office space to seat the thousands of executives they were hiring—all over the world. Their stocks soared with the surge in revenues, as Lucent became one of the most valuable companies in the world, and Nortel made up one-third of the Canadian market's entire valuation. Nokia's stock comprised two-thirds of the total value of Finland's Helsinki Stock Exchange.

The stock market gains fed further success. Equipment makers used shares of their wildly overvalued stock as currency to acquire control of venture-backed companies and smaller public companies. This technique kept their cash free for other purposes, such as new technology investments— or a business practice that would push the industry to the edge of the abyss, and then over it.

When a company is seeking to grow market share and boost top-line revenues, one common practice is to "buy business." This typically means selling at a low margin, or maybe for no profit at all, in order to beat your competitors. The problem with this technique is that it actually costs money to do it. Selling at a low margin means your revenue may grow, but your profit drops and you have less money for investing and for paying bonuses to executives. In short, the problem with buying business this way is that the cost of doing it is *real*.

But there was another approach that was . . . less real.

The huge national service providers like Verizon had ample cash flow and didn't have trouble coming up with the money to buy new gear. They negotiated the best price they could, then paid cash. But for the smaller start-up carriers, it was a very different story. They didn't care so much about the price. Their goal was growth, not profitability, and they needed to get their networks up and running as fast as they could, cost be damned. Unlike the big carriers, their cash was in short supply, coming from investors until they could build significant customer bases and generate real revenue.

As a result, the equipment vendors' sales teams developed a deal sweetener called "vendor financing." Equipment makers would quote a network build-out for the new company—say a complete turnkey switching system, wireless gear, and transmission equipment for $100 million. Maybe the price was higher than what would be offered to the larger incumbent carriers, but vendor financing provided something important: the equipment came with a loan to buy it. The vendor would deliver the gear for a payment of perhaps $60 million and a loan of $40 million. Or maybe—and this became the norm in the industry—they would offer it for a payment of zero. The vendor would deliver the $100 million network, take no payment at all from the start-up, put the $100 million credit on the company's balance sheet, and book the sale as revenue.

This had the result of burying the lack of any payment on the sale somewhere in the balance sheet, not on the income statement, where analysts and investors would see and understand what was happening. By listing the "loan" on the balance sheet, it would appear as an asset of the company, a credit that would be paid back in the future. Convenient and less "real" than acknowledging that they hadn't received any payment for all these booked sales.

Needless to say, the small start-ups liked the approach too, but soon getting the gear for free wasn't enough. They wanted more. Cheaper than free. Vendors saw this desire as reasonable and began to do something inconceivable. They began *paying* new customers to take their gear. A pre-revenue start-up—a company that had no paying customers yet—might buy $100 million of network equipment, pay nothing, and receive the network equipment *plus* $20 million in cash from the vendor to spend on salaries, marketing, acquisitions. Executive bonuses. What could go wrong?

Banks are experts at lending—they know how to evaluate the likelihood a loan will be paid back and how to secure payments with collateral. As Frank D'Amelio, who put an end to the practice when he was made CFO of Lucent Technologies, now describes it, "The reason the start-ups were coming to a supplier like us for financing was because they couldn't get financing from a traditional lending institution. That was a big red flag." So why *weren't* the banks making these loans to the new carriers? Because *banks are experts at lending—they know how to evaluate the likelihood a loan will be paid back and how to secure payments with collateral.*

The banks wouldn't have *touched* these deals.

The equipment makers like Lucent had some of the smartest people in the world working for them. Their labs were filled with PhD physicists who graduated first in their class at MIT. They were literally staffed with Nobel Prize winners who understood networks and technology better than anyone else in the world.

But they didn't understand the subtleties of lending, and they weren't staffed to evaluate the risks and pursue repayments. Most of these companies had limited treasury teams responsible for managing their employers' *own* borrowing and financial needs, and they were already stretched thin trying to keep up with their own breakneck growth. There was no capacity to evaluate loans to outside companies with no credit history, and few assets to serve as collateral beyond the ones the vendors were delivering to them. The salesmen kept making the deals, and the financial officers began sweating as the "credits" built up.

In 2001, Lucent, with $21 billion in revenue after spinning off some divisions the prior year, financed nearly $8 billion of its own equipment sales.

ROLLING IN THE GREEN

If there's one investment that makes vendor financing look prudent by comparison and showcases the excesses of the era for the equipment makers, it's the "business development and marketing" initiative that was a pet project of Lucent's CEO. It involved an eighteen-hole championship golf course under development at the site of the former James Brady estate in Gladstone, New Jersey. Thirty-six holes, actually, as it included the country's only USGA-rated par-three course. Rich McGinn, CEO of Lucent Technologies, had a marketing plan: Lucent would build the courses, refurbish the sixty-four-room former mansion of James Brady, the man whose son served as President Reagan's treasury secretary, and invite eighteen CEOs to join as members and co-owners of the most exclusive golf club in the world. All in the name of business, of course, as these CEOs would represent the largest buyers of Lucent's networking gear. The heads of AT&T, British Telecom, Verizon, and others all would enjoy the generosity of Lucent and, no doubt, keep that generosity in mind when it came time to deploy a $100 million local-access network or select a vendor for the next wireless build-out.

For a vendor whose only customer used to be its parent company, this was marketing. The days were gone when Western Electric, the former subsidiary of AT&T (and Lucent's predecessor) could make communications gear according to their own specs and sell it no bid into AT&T's network. Now, Lucent was faced with competing vendors like Nortel, Alcatel, and Ericsson, well-funded companies with good technology.

But Lucent also had multiple customers they could now court, and the company was creating a presence in the market, backed by Bell Labs technology and spearheaded by an aggressive sales force recruited mostly from outside of the company's own ranks. That sales force had been killing it, beating ambitious growth targets in the Americas and cracking open new markets in Asia. Revenues were heading to a peak of nearly $40 billion a year, and employee count was approaching 160,000. The market cap of Lucent soared past $250 billion, as the company became the most widely held stock on the NYSE.

The company's facilities team had tried to explain the problems with the course, both from a cash perspective—the course's costs had ballooned to over $40 million—and in terms of the appearance to shareholders, customers, and regulators. They knew that the days of 20 percent year-over-year growth wouldn't last forever, and they suspected the company would soon face its first-ever belt-tightening. The efforts to sign up other companies as club members/co-owners was also not going well, leaving Lucent on the hook for an increasing liability. Some customers had suggested describing the facility as a conference center that *happened to include a golf course*; that might make their own boards view the "investment" more favorably. But the effort was not making much headway.

Despite the concerns, the team had their orders to proceed with the development. Things were going to get worse for the company, but they couldn't know how bad they would get. By the time the fairways were completed a couple years later, Lucent would stop watering the lawn around the executive parking lot of their headquarters in a desperate attempt to conserve cash and cover payroll.

That year, as Lucent was pouring money into the golf course, Huawei was using its money to build an R&D center in Stockholm, having just launched its first international R&D facility in Bangalore, India. By 2001, as Huawei opened four R&D centers in the United States, Lucent was exploring plans to unload the golf course to an investor group for a million dollars and a note.

IN THE DARK

It was only a matter of time. When the dot-com bubble burst, fortunes were wiped out and absurd valuations brought to zero. The telecommunications services that were going to enable the internet explosion evaporated, for the time, at least, and with them the need to build network capacity to carry those services. Dozens of start-up service providers crashed, declaring bankruptcy before they could uncrate the switches and turn up the new cell towers, and when they did they left their equipment vendors on the hook for their massive build-outs. D'Amelio says the impact was immediate and terrible: "When the new competitive carriers went belly up, we had to write it all off." Billions.

By 2001, Lucent shares would be approaching penny-stock level and the company would reduce its workforce by 80 percent to fewer than thirty thousand employees. The telecom equipment manufacturing sector would lose over one million jobs and $1 trillion in market value that year. Much of the precipitous decline suffered by the industry could be attributed to the collapse and consolidation of their customer base when the internet bubble burst, but there was plenty of blame to assign to bad management decisions and poor product planning.

One Monday, after reviewing the cash crisis he had inherited, D'Amelio met with Lucent's global head of facilities in his office.

"I want the maintenance team to unscrew every fourth light bulb in the offices," D'Amelio said.

"Which offices?" he asked.

"All of them. Starting with headquarters. And don't wait till the evening shift to do it. Do it during the day, when people are here."

The facilities VP was confused. "Why are we doing this? And why during the day? Do you realize what you're doing?"

D'Amelio explained: "I know exactly what I'm doing. We need to shock the system. I want people to see it and think about the mess we're in. We need to stop spending money. We need to preserve cash." D'Amelio paused for a minute, then continued, "Also, stop watering the lawns."

The message was sent, and more material measures were also taken, as D'Amelio cut the dividend, mortgaged the company's main buildings, and factored receivables to banks. The emergency measures stabilized the bleeding, but the worst was yet to come.

For the largest, established carriers, the pressure to upgrade their networks abated as the new competitors failed. And how they failed, almost to the last one! WorldCom became the biggest bankruptcy in history. Bankruptcies swept the sector, and with them purchase orders from start-ups went to zero. Capital spending by the large, surviving carriers plummeted. Analysts estimate that spending at the largest carriers dropped from $78 billion in 2001 to $49 billion in 2002.

One story is particularly illustrative of the chaos during this collapse. On April 18, 2001, Winstar Communications declared bankruptcy, unable to repay bank loans that had come due. In their filings, they blamed Lucent for breaching an agreement to make the latest vendor-financing payment of $90 million to them on March 30 of that year, which Winstar planned to use to pay back bank loans. To be clear, Lucent, which had delivered equipment to Winstar but not yet taken any payment, was late on its transfer of $90 million in vendor-financing cash payments to Winstar. And Winstar planned to use some of that money to make the latest payment *back to Lucent* on the $700 million Lucent had loaned Winstar to buy the equipment for their network.

Upon Lucent's failure to make the payment on time, Winstar promptly sued Lucent for $10 billion for breach of contract.

If it seems it can't get crazier, some months after the dust settled on the bankruptcy filings, an agent brought a proposal to the leadership at Lucent. Winstar had a warehouse full of brand-new switches that they had purchased from Lucent for one million dollars each, still sitting in the crates. The cost for Lucent to manufacture and deliver these switches was about $600K apiece, but Winstar would offer them for repurchase by Lucent for the bargain price of $500K each, half of what they commanded on the open market and—more importantly—less than the manufacturing cost to Lucent. It was a bargain. Lucent agreed to consider the proposal.

Given that Lucent had built the switches, sold them to Winstar, paid Winstar over $100 million dollars to finance operations, and collected no money on the sale, they decided that they did not like the optics of buying them back.

Lucent passed on the offer.

10

Deal or No Deal

"How did you go bankrupt?"
"Two ways. Gradually, then suddenly."
—Ernest Hemingway, *The Sun Also Rises*

Just days after the Winstar bankruptcy and barely a year after Lucent had announced record profits, Lucent realized it needed to find a buyer or face bankruptcy itself.

BULL IN A CHINA SHOP

Thursday, the week before Memorial Day, 2001, found Frank D'Amelio on a flight to Paris. D'Amelio had kept his air travel to a minimum in the past. He didn't like flying. For that matter, he wasn't that comfortable in cars either, preferring to ride in the front seat of his company car and even taxicabs to avoid carsickness. But he had just been named CFO of Lucent, and his new boss had asked him to join the team to negotiate a merger with rival Alcatel. The proposed deal was so secret he had only found out about it a week earlier, upon his promotion. D'Amelio spent the flight catching up on the deal terms that had been developed with the previous CFO.

The negotiations were hosted about an hour from Paris in what looked to the Lucent team like a castle, used by Alcatel's top officers for off-site retreats and meetings that couldn't risk discovery. This was supposed to be a merger of equals, but D'Amelio had the feeling that Alcatel's leaders were treating this like a hostile takeover without the premium, and Lucent was the prey.

That feeling intensified as he listened to the First Boston banker leading the deal for Alcatel, George Boutros, who had a reputation as a "yeller." A *Wired* magazine article published days before the meeting described him: "He doesn't just want to win. He wants to win and make you look stupid. He roughs up the executives from the other side so badly that they want him on their side next time."

D'Amelio was aware of Boutros's reputation. His predecessor as CFO at Lucent had been pushed by Boutros into paying over $20 billion (in stock, of course) to acquire internet switch-maker Ascend. It hadn't worked out well, at least not for Lucent. Boutros would find D'Amelio a bit harder to rough up.

Of all the deals Boutros had led, this would be among the most important. Alcatel knew that Lucent urgently needed to close a deal, and even though they didn't realize the extent of the crunch Lucent was in, they planned to squeeze them hard, describing strict terms to be applied from now until the deal received approval from the regulatory and antitrust authorities. It's not known if Alcatel had shared with Boutros their own urgency about closing a deal quickly, but that would be exposed shortly.

As Boutros read out the terms and conditions, D'Amelio began to shift in his seat, feeling his blood pressure rise. The lawyers sitting around the edge of the room glanced at each other uneasily as they saw the muscular executive squeezing his hands into fists. D'Amelio had been a football star when he was younger, recruited to play linebacker for the University of Florida, and he still had the build of an athlete who enjoyed contact sports. His blunt demeanor could be seen as menacing, at least for people who were trying to pull something on him. His demanding schedule had been making it harder to get to the gym as often as he wanted, but he still started every day with a hundred push-ups and a hundred sit-ups.

D'Amelio listened to the growing catalog of restrictions and finally waved Boutros off. He was not one to be steamrolled by anybody, including a Wall Street bully.

"Guys, we have to be able to compete until this deal closes, because it's possible it won't," he said. "I can't agree to these terms." The Alcatel executives shook their heads and whispered quietly to each other. There was no need to lower their voices; no one on the Lucent side spoke French.

At the next break, D'Amelio slipped into a conference room and called Lucent chairman and CEO Henry Schacht, who had replaced Rich McGinn, in Murray Hill.

"Henry, listen," he said. "All this stuff that you told me we agreed to, it's all being revisited. As your CFO, I don't know how we can do this deal!"

Schacht was taken aback as D'Amelio walked him through the terms of the deal, including elements that Schacht had agreed to with Alcatel chairman Serge Tchuruk, then said, "That's not what Serge and I agreed to . . ."

D'Amelio shook his head. "What you and Serge agreed to is not what I'm hearing in this room. Now Henry, I wasn't part of this, but . . . it's a problem. A big problem."

Schacht hung up the phone and called to his secretary, "Have them get the plane ready. This morning, for Paris."

Later, Lucent's head of corporate security would curse about being left out of the loop on the super-secret negotiations. "I can guarantee you they had bugs in every room," he said, "taps on every phone and fax machine. I should have been there watching over our guys."

Schacht arrived in Paris Friday night, and on Saturday morning he and Tchuruk took a walk in the garden at the castle. Beyond the two chairmen, painted on a wall of the castle, a massive mural of Napoleon looked down on the scene as his countrymen engineered the takeover of Bell Labs, the crown jewel of American innovation.

A group of Lucent executives followed Schacht and Tchuruk at a distance. The body language between the two men looked good, but the Lucent folks were not close enough to hear what was being said. At the end of the walk, the chairmen shook hands and smiled warmly, then broke to meet with their respective teams, where they let them know the deal was back on track. But by Saturday night, it had fallen apart again, and by Sunday morning the deal was off, with the Lucent team on a plane headed back to the Morristown, New Jersey, airport.

Lucent's catering department had already ordered French pastries, croissants, and café au lait for a big press announcement at the St. Regis in New York. The order would have to be canceled.

The deposit was not refundable.

TROUBLES SPREADING

Memorial Day Monday, the leadership team took advantage of the empty offices and assembled again in the CEO's anteroom in Murray Hill to figure out what to do. At the end of the meeting, Schacht sent D'Amelio off to review their company's financials and come back with an assessment.

"With the Alcatel deal off we have to make this on our own," said Schacht. "Tell us how bad it is."

D'Amelio spent the next few days holed up with his treasurer, reviewing the business's financials. A company of Lucent's size, with blue-chip customers who pay their bills on time and triple-A access to debt markets, typically didn't worry about cash. But what D'Amelio found as he reviewed the books—many of them for the first time—stunned him. He called the leadership team back together.

"We are hemorrhaging cash," he told them. "We've got a few months, guys, and then we're out of money."

It was only a few weeks later that D'Amelio opened the *Wall Street Journal* and saw a surprising headline. Alcatel was announcing a three billion euro write-down, a downward forecast for revenues, and a clean out of senior managers. It hadn't come up in the negotiations, and it should have. But D'Amelio had spotted anomalies in their financials when he reviewed them at the table in Paris—he could read a balance sheet like he could read a sports page—and he had reported back to Schacht that there was something else going on at Alcatel that didn't make sense.

Lucent was being hurt by troubles in the US market, while also experiencing price pressure around the world from Huawei, which was low bidding every contract, even when they knew they couldn't win. It appeared that Alcatel, which did a large portion of their business in developing countries across Africa and Asia, was also taking a beating from the new kid on the block. Alcatel, Lucent's would-be savior, was in rough shape themselves, losing share in the developing markets and seeing their own pricing destroyed by a competitor that seemed to have no bottom to their bids.

If the deal had closed, Alcatel's losses were the type that would have been easily rolled into the confusion of merging two massive companies' books. It was a sign that things weren't just bad for the American vendors.

11

The Sincerest Form of Flattery

> The companies in China copied our technology, but we
> were in such a fast-changing sector that by the time they
> brought it to market it was obsolete. They finally gave up.
> —Mike Tessler, CEO of BroadSoft

While the Western companies were on a roller coaster of highs and lows in the late 1990s and early 2000s, Huawei consistently made strides. A little over a year after introducing their first analog switch, Huawei had rolled out and started selling a digital switch. It was an incredible achievement, and Frank D'Amelio had found out the hard way when he was visiting Beijing to tour a Lucent joint-venture switching facility back in 1997.

It was a few years before D'Amelio would be named CFO, and he had just been assigned to help lead Lucent's switching equipment division. His local team had something to show him: An engineer led him into the switching center where the latest Huawei switch, the C&C08, was laid out next to Bell Labs' flagship 5E switch, the pride of America's telecom network. The engineer brought him over to see the switches, placed side by side. He opened the back panels to reveal the inner workings and looked expectantly at D'Amelio.

"*Unbelievable*," D'Amelio said, grimacing.

They were identical. The layout, the framing. Even the cabinets were the same color. Each switch had a large bundle of wires connecting the two main sections to each other. The colors of the wires were identical. As the engineer closed the cabinets, D'Amelio let out a laugh. Yes, the logos were different, but the ventilation holes in the Huawei switch had been drilled in the exact same pattern as the Bell Labs switch.

"It's the same, isn't it?" he asked.

"Not exactly, Mr. D'Amelio," the engineer replied. "It's almost identical to our switch, but they've made some changes. It's more suited to markets with cheap labor—there's less automation. It runs on less power, for areas with weaker infrastructure support."

Somehow, Huawei had built in months what it took Bell Labs decades to develop. *And it looked just like what Bell Labs had developed.* The Huawei engineers, D'Amelio was informed, were *proud* of how closely their switch resembled Bell Labs'. When D'Amelio later told his boss, switching president Bob Holder, about the appearance of the Huawei switch, Holder shook his head, then paused for a moment.

"Dollars to donuts the software looks the same too," he said.

A few years later, in 2002, Bell Labs performed a technical assessment on the next version of the Huawei switch, which showed that they had made further leaps. According to the report, "The second-generation hardware was more Softswitch like." That meant it looked less like an older voice network switch and more like an internet protocol switch, the kind D'Amelio was driving his own team to develop. The Chinese equipment makers were catching up and now passing the leaders.

How could this have happened?!

OR JUST PICK ALL THREE

It's a fact of technology development that there is one limitation that even scale can't overcome: Research and development projects require *time*, no matter how many people you throw at them. As the trope goes, it takes nine months for a woman to have a baby, but nine women can't have a baby in one month.

Unless you send them out to steal someone else's baby.

In developing a product, managers—no matter their company or resources—have to make a choice. Fast, Good, or Cheap: pick two. There are always trade-offs in development. If there's a rush to get something to market, you can crank it out quickly and cheaply, but it won't be very good. You can get a good solution, quickly, if you pay a premium. Or you can find cheap developers to do the work well, but you'll need to spend a longer time doing it and ironing out the bugs.

The point is, you can't get a new product developed well, in a hurry, on the cheap. At least, you can't do it without committing a felony. Several criminal and civil cases have alleged that Huawei engaged in the illegal appropriation of intellectual property from competitors, in effect bringing itself up to speed impossibly fast by procuring technology that other companies had spent years developing.

This is not a simple matter of "reverse engineering," an accusation often thrown against Huawei. It's a hollow claim. In the rough and tumble of business, there is a lot of leeway to compete aggressively and remain within the law. Reverse engineering is a legitimate technique, and it typically includes disassembling a competitor's product to see what makes it tick, cracking the box open to learn how the other guy is doing it, and figuring out how you can respond. Anything you learn that isn't protected by law, such as patents, is yours to copy. Exceptions exist, for example, in the case of someone who acquires the equipment under a nondisclosure agreement. Huawei certainly reverse engineers other products, as do all their competitors.

For the software portion, competitors can't just crack open the case. The magic, called "source code," is not reviewable by anyone except the owner who possesses the key to view it. But again, this is for those operating within the law.

COINCIDENCE

Cisco, the California-based internet switching-gear giant and one of the United States' leading technology exporters, had encountered its own problems with Huawei allegedly copying proprietary software and user's manuals for a key Cisco router. Huawei was having trouble getting a competing router to market fast enough and was trying to work through quality problems that were delaying release. Remember Fast, Good, or Cheap? According to Cisco, Huawei decided the best way to get a good router to market quickly and cheaply was to appropriate one that Cisco had already spent time and money making good.

The court filings from the US District Court for the Eastern District of Texas present testimony from Chad Reynolds, a Huawei employee based in Plano, Texas. He stated, "I believe that Futurewei (Huawei's wholly owned US subsidiary) is in possession of Cisco's source code . . ." In his testimony, Reynolds described how Huawei could not ship its clones of Cisco routers because "they contained too many problems that were the same as Cisco routers had . . . The presence of many common bugs between Cisco and Huawei would indicate copying." In addition, Cisco argued that the manuals for the Huawei routers contained typos that were identical to those in the copyrighted Cisco manual.

According to the *Wall Street Journal*, Cisco's general counsel flew to Shenzhen to present the damning evidence to Huawei's leadership, including identical comments written in the code and verbatim typos in the Huawei user manual. Chairman Ren listened to the presentation impassively before responding: "Coincidence."

Huawei settled with Cisco, agreeing to change its products but without admitting fault, after the Neutral Expert appointed by the judge found that "it must be concluded that Huawei misappropriated this code." But the damage was done; within two years they had crushed Cisco's dominant market share in China with their clone routers.

THE ENEMY WITHIN

Lucent learned that as concerned as they were about outside companies stealing their products, they also faced a threat from within their own labs.

In 2000, internet phone calling, known in the industry as "IP voice," began to emerge as an option, and cable operators across the United States started to develop their own business plans to add digital voice services to their television and internet offerings. Lucent was eager to supply them with gear to enable these new services and introduced a product called PathStar, which would allow cable TV providers to provide voice calling similar to traditional phone companies. The product had landed some major early customers, including Comcast and Time Warner, and Lucent expected this product would develop into a major market. If Lucent could secure a lead, no amount of money could help another company immediately match the offer.

But during routine oversight of the company's systems, corporate security at Lucent determined that three foreign nationals working for Lucent had downloaded and stolen the PathStar source code, the guts of the system. In a superseding indictment dated April 11, 2002, Chris Christie, then US attorney for the District of New Jersey, laid out the allegations: The three had "conspired to steal and transfer the software and hardware" to a Beijing-based telecom company. After obtaining a search warrant for their homes, "FBI agents seized large quantities of the component parts of the PathStar Access Server, both software and hardware, as well as schematic drawings and other technical documents related to the PathStar Access Server . . ."

And here's where, as a Lucent board member confides, "It began to get weird."

The prosecution did not go well, and it wasn't for lack of compelling evidence.

"Three individuals were caught downloading the code and the company turned the case over to the FBI," says former Lucent CEO Patricia Russo. She was not one to be pushed around. Despite working in a traditionally male-dominated industry, she had risen to become the leader of one of the most respected companies in the sector and would later be named

Chairwoman of Hewlett Packard Enterprise and take seats on the boards of General Motors and Merck. Tall, dark-haired, and striking, Russo could drive a golf ball farther than her male colleagues and was not shy about doing what she thought needed to be done. But even she could not anticipate what happened next. "It became apparent over time that the Chinese were angry at Lucent for taking action against individuals for the theft of our intellectual property. We felt that *we* were in the penalty box in terms of our business with China."

Lucent was selling hundreds of millions of dollars into China Unicom and China Telecom, and that business fell off considerably; the company learned through discrete contacts that Lucent was suffering as a result of the FBI investigation. Russo recalls meetings in China where the issue would come up subtly and she would have to navigate through that. "I had the impression they wanted an apology for turning the case over to the FBI," she says. "I remember leaving one of those meetings and saying to my colleague, 'Don't they realize *they* broke the law?!' But China had a huge market."

The consequences of pursuing the case would have been the loss of hundreds of millions—perhaps billions—of dollars in sales during a time when Lucent was fighting for its life. Any CEO representing the interests of her stockholders would have faced the stark choice: Do we support prosecution of these three low-level engineers (two, really, as one had already jumped bail and was believed to have fled to China where the odds of a successful extradition were low), or do we settle this quietly and get back to business? Lucent chose the latter, eventually settling for a small payment from the accused and agreeing to drop the matter.

"If you wanted to do business in China, there wasn't a lot of complaining," Russo says.

There was, however, a bit of irony in that Lucent ended development of the PathStar line before the prosecution was even resolved, citing budget cuts and, less publicly, problems with the same source code the three engineers had downloaded. The original developers hadn't followed the best practices in writing the software, and estimates were that it would have taken the Chinese "entrepreneurs" years to sort it out and turn it into a sellable product, by which time it would have been long surpassed by competing solutions.

While all countries experience malfeasance by their citizens abroad, the odd thing here is how the Chinese officials and executives closed ranks. If it would have cost Lucent a great deal of lost business to prosecute these accused thieves, it was costing their Chinese customers—ostensibly unconnected to the bad actors—a price to deny themselves the equipment they really wanted to buy from Lucent. Why were other Chinese companies and the government willing to bear a cost to their own goals in order to protect three accused thieves who supposedly didn't even work for them?

12

A Foothold in Europe

If you want something in the worst way,
that's exactly how you'll get it.
—Unknown

In 2001, as the wheels came off the suppliers across North America and Europe, Huawei took their newfound experience in international bidding and went on a tear, winning business around the world and landing their first major contract in a developed country. Their strategy for winning consisted of rock-bottom prices and serviceable, if not leading edge, products. According to the companies that were trying to compete with them, Huawei's approach also included bundling, vendor financing, and, as claimed in multiple civil and criminal cases, more aggressive tactics.

THE FIRST BIG WIN

The deal that cracked open Europe for Huawei involved one of the world's largest telephone companies, run by a scrappy iconoclast. Ben Verwaayen prided himself on being demanding and not just going along with the flow. More than one corporate board had learned this the hard way after inviting the senior telecom exec to join them, only to find him uninterested in rubber-stamping whatever matters were brought to them.

Compact, intense, with round professorial glasses, his contrarian nature started long before he had the currency that came with being a Fortune 50 CEO. When he graduated college in the Netherlands, Verwaayen decided to carry out his military service requirement by enlisting in the Dutch army, where soldiers are allowed to form and join unions. Verwaayen felt the existing union was too politically extreme and promptly created the new General Association of Dutch Soldiers, a union that remained in place until the end of the Netherlands' forced conscription, decades later.

His next job was with a small insurance company owned by ITT, the conglomerate that years earlier had come under fire for their role in working with the CIA to depose Chilean president Salvador Allende. Verwaayen caught the attention of the PR department when he proposed that the company become more open about ITT's history and conflicts. To their credit, the company saw him as a high-energy manager who was not constrained by normal views of propriety. He was promoted to run PR, then moved to the telecom division, and, by thirty-six, he was CEO of the Dutch national phone company. In 2003, at the age of fifty-one, he was named CEO of British Telecom, one of the largest carriers in the world.

Verwaayen had no intention of being a caretaker, and in the spring of 2005 he launched an audacious bet-the-company gambit to remake the United Kingdom's national network.

Early on in his tenure, Verwaayen sat at a bar with Matt Bross, the chief technology officer he had just brought on, and grabbed a cocktail napkin to sketch his vision of the UK's new network. Verwaayen believed that not only was broadband the future, but that a radically different approach would be required by the network in a few years. Not all telco executives around the world had this understanding, but Verwaayen was right, and he was willing to put £10 billion behind that bet and do it at breakneck speed. He also needed a partner who was willing to look outside the usual stable of vendors to achieve their goals, and saw a kindred spirit in Bross—or "Mad Matt," as he called him.

"Matt translated that napkin into a really serious redesign of networks," Verwaayen says, "and he was the absolute architect of what later became a kind of seamless networking based on software integration. He looked to Huawei as an enabler of that, at the time when nobody else did." Verwaayen says he left the vendor selection to Bross and, in April, British Telecom announced that Huawei would be a key supplier to the "21CN," BT's new 21st Century Network.

Bringing Huawei in on this massive contract—one of the largest in the world to that date—was a shocking moment for the industry. Marconi, a British company Verwaayen describes as the "national pride supplier," was one of Huawei's main competitors and the presumed favorite to win the edge portion of the network. But the BT team wasn't impressed with their ability to execute, and Verwaayen felt that their solution was not in the same league as Huawei's.

"There was no competition from the technology side," he says.

Not long after the announcement of the Chinese vendor, Verwaayen's phone rang. It was Mike Parton, CEO of Marconi. He couldn't acknowledge what both of them knew: Marconi was unable to match the design, prices, or support services that Huawei was proposing. But they both also knew the consequences of losing the bid.

"Ben, you know what this loss will do to us, don't you?" Parton pleaded. He argued that it would be devastating to his company, the last major UK-based vendor of telecom network gear.

It was. The hundred-year-old firm, which had barely survived the 2001 telecom meltdown and was already facing a death spiral, collapsed in the face of this low bid by Huawei, losing 40 percent of its stock value that day. By October 2005, most of the company was scooped up by Ericsson, with remnants going to an American private equity firm a few months later.

The other established Western vendors were also caught off guard by BT's move. They had worked their relationships with the CEOs of the biggest carriers and that, along with government pressure, had kept the Chinese company out of all of the large national networks, effectively consigning Huawei to second- and third-tier service providers. But Verwaayen was never one to play along, and now Huawei was helping build the most sophisticated network in one of the world's most advanced countries.

Pat Russo, CEO of the recently merged Alcatel-Lucent, who was another favorite for the BT contract, looks back on the award to Huawei as a "gut punch." It seemed impossible to compete with Huawei, whose pursuit of the contract was relentless. "They had five engineers to every one of ours," she laments.

Verwaayen understood the importance of a healthy vendor community to his company. He says, "I was concerned enough for the viability of a competitive landscape that I said, okay, it's great to have Huawei at the edge, but we can't also have them at the core of the network." Conventional wisdom is that the core of the network is the "crown jewels"—that the core is the part of the network that most needs to be kept secure. In fact, evolving technology had already made even the edge a source of vulnerability, and BT would be shocked to see the consequences of this a few years later.

His vendors' hurt feelings aside, Verwaayen and BT were in no condition to serve as benefactors to the equipment industry. The company was financially troubled when he took it over, and building a state-of-the-art network was only part of the rescue equation.

"BT was in bad shape," says Verwaayen. "Because of the buildup of debt, we had to sell off all the future-related businesses like mobile—this was a fixed-line only business. We had to do something that was extraordinary and would also revolutionize our costs. From every single angle that we looked at it, the concept that Huawei brought to the party was massively better than other companies would offer."

CHATTER

In 2010, less than five years after 21CN went live, sources informed the Australian Broadcasting Corporation (the ABC) that some of the core switches installed in the BT network by Huawei were found to be doing a lot of "chattering"—industry jargon for when a network element is transmitting more information than the traffic should require. That's a red flag for companies concerned about their data being exfiltrated to unknown end points. In short, BT and the British security services were alarmed that the Huawei switches were sending information off to an unknown recipient. Senior officials from both parties put themselves on a plane to Shenzhen in 2011 to sit down in Huawei's headquarters and discuss their concerns.

Although further details surrounding this claim have not been made public, it's clear how they felt after the meeting. The ABC reported that BT chose to remove and replace the identified Huawei switches at considerable expense. That was hardly the end of all Huawei gear in the network, however.

Meanwhile, the British government commissioned a review, which was overseen by Malcolm Rifkind, chairman of the Intelligence and Security Committee. The committee sought to understand how the government's ministers evaluated BT's plan before allowing the Chinese vendor to deploy its gear into the United Kingdom's national phone network. What was the review process? The answer was surprising.

There was none.

The committee determined that government bureaucrats were concerned about getting in the way of the deal, given the trade implications for the UK and the potential impact on British-Chinese diplomatic relations. They decided not to brief the ministers at all until after the deal was inked.

While accurately recognizing the failures in oversight, the committee then made an error of its own. They concluded that risk was already present in any network that contained gear made in China, even if that gear was made by a British, American, or Swedish company, and there was no point in keeping any given country's equipment out of the network.

This facile perspective misses one of the key dangers of having a non-trusted entity deliver its own gear into a network. When a company like Ericsson manufactures a switch in China, they are motivated to ensure the integrity of the software and hardware, which includes scrutinizing their own supply chain and independently ensuring that the factory is manufacturing exactly to specifications. More importantly, once the gear is deployed, it is Ericsson engineers who retain control of the device during its lifetime in the network, monitoring it, updating it, and controlling who has access.

But when the equipment is not just manufactured in a nontrusted country but designed, manufactured, validated, updated, and monitored by a national champion of that nontrusted country, you lose the extra layers of scrutiny and ultimately turn over security to the very party in question. The difference is enormous.

FOX WATCHING THE HEN HOUSE

Not long after BT's 21CN was up and running, the British government identified the security risks of using Huawei's equipment in critical national infrastructure and established a Huawei Cyber Security Evaluation Centre (HCSEC) Oversight Board to inspect Huawei gear, requiring Huawei to make its software code available for inspection. Huawei dutifully did so, but many experts consider this "security theater" for several reasons.

First, communication equipment is complex, with some elements bearing millions of lines of software code. The architecture and design vary from company to company, and even the best-trained scientists can't examine a new company's product and fully understand what they are looking at.

Second, the oversight board is only able to look at what is sent to them. A national network may require hundreds of thousands of components to be deployed, and the review team can only evaluate the samples that they are given. This doesn't represent everything that is going into the network.

Third, networks are not static. The hardware constantly changes as problems are discovered or incremental improvements are made. A device deployed into a network on day one may undergo dozens of adjustments and alterations over its service life. More troublesome is the need to constantly update software, not just for performance enhancements but for security needs. As gaps and flaws appear in a network's security—and this is a constant, universal fact of life in all networks—the vendor must make patches.

Patches are a constant issue with networks. And when a "zero day exploit" is discovered, which means the first moment the good guys learn they are vulnerable to an exploitation, there is a mad scramble to develop and deploy the patch. That emergency security patch is not something that can be sent to a "review committee" so they can convene a team of experts and explore the patch code to determine whether it, too, presents a risk. That is an all-hands fire drill that assumes absolute trust if not absolute efficiency.

There is one other limitation to using the HCSEC to vet Huawei gear, and it's a big one. The center is paid for, owned, and, to a great extent, managed by Huawei. While this structure doesn't necessarily mean the center is *completely* under the control of Huawei, the danger is obvious. Enough so that the UK retains Ernst & Young every year to audit the operations and issue an opinion on HCSEC's ability to operate independently of Huawei. In the 2020 report, EY concluded that there were "no major concerns."

As for the "minor" concerns? The report states, "Overall, the Oversight Board can only provide limited assurance that all risks to UK national security from Huawei's involvement in the UK's critical networks can be sufficiently mitigated . . ."

Did this breaking news cause panic among the protectors of the UK's networks?

Well, it wasn't exactly breaking. Similar warnings had already been issued: EY notes that "as highlighted in previous reports, HCSEC's work has continued to identify concerning issues in Huawei's approach . . ." and states that, "Limited progress has been made by Huawei in the remediation of the issues reported last year . . ."

So even when problems are found, as they have been for years by the HCSEC and other organizations, that doesn't mean the problem is remedied. It is simply identified.

The error here may be in the assessment of the intent and character of the company supplying the equipment. When dealing with complex, proprietary systems, it's not unusual for a customer to demand that the vendor set up processes to make sure everything works as required. Self-policing can make sense. The creator and supplier of the gear is best suited, after all, to evaluate performance and identify problems and find solutions. But this model is only appropriate when the relationship between the parties is one of trust and faith—we're all in this together, and no one wants to see something bad happen to the customer.

When the concern lies with the *intent* of the vendor, such a model is a farce.

If residents of an apartment building were trying to secure their apartment against burglars, they could have their landlord put in a high-quality lock. They could make sure the door was heavy and hard to smash in. But if the person trying to burgle them was their *landlord*, none of this would matter. He has the key. Not only would all the measures be pointless, but the tenants would have no way of knowing if they had been burgled: no broken lock, no smashed doorframe. Just a violated apartment.

Huawei was deployed into the network under the scrutiny of BT CTO Matt Bross and service was turned up, riding on billions of dollars of network equipment. Millions of Britons were able to enjoy high-speed internet at rock-bottom prices, but a known threat of unknown consequence was injected into the nation's network, with remediation options potentially costing billions. No level of scrutiny would make this problem go away. And the bill would come due.

13

Sizing Up Huawei and the Global Marketplace

For less than the cost of an aircraft carrier, China's subsidies
to Huawei wiped out the greatest source of technology
innovation in America—and perhaps the world.
—Anonymous, former executive at Lucent/Bell Labs

How were well-meaning companies so vulnerable to making what seemed to be bad choices when it came to national security? The problem BT and others faced was that the benefits of choosing Huawei were real, measurable, immediate, and accrued to the buyer. The risks of making that choice were vague, unmeasurable, distant, and accrued to others. This explains why Huawei's pricing was so hard to resist.

But it doesn't explain how their pricing could be *so much* lower than anyone else in the industry. There are cases where a vendor may have a special product that allows it to charge a premium—while that advantage lasts—and situations where a company comes up with a unique approach that enables significant costs savings—again, while that approach remains unique. But the range of prices submitted in response to any request for proposal is generally pretty narrow.

Within a geographic market, everyone pays their scientists about the same salaries. Salespeople work off similar commissions, based on the selling region not the home country, and they move frequently between companies (providing additional insight on competing price ranges). Rental rates and electricity costs don't care who's in the building, and costs for raw components are well known and fairly level, with those components typically sourced from common vendors.

In the global marketplace, companies take advantage of the strengths of each market: software from India, hardware and assembly in China, systems integration and network engineering from Europe and the United States. It is a myth that Chinese products have a cost advantage because of the cheap labor in China; the factories of Nokia and Ericsson are in Shenzhen down the street from those of Huawei and ZTE. The result is a fairly flat cost range between vendors, with pricing moving up a little if a vendor has a unique capability and down when they are hungry to win the business. The calculus comes down to how *hungry* a vendor is to win a deal. As a result, bids tend to vary mostly by each company's willingness to give up margin to win.

This raises important questions about the prices that Huawei used to grow its share around the world. In many markets, when a vendor has a particularly strong desire to secure a contract, prices can approach what is technically considered "dumping," a dirty-sounding word but not necessarily an illegal practice, where a project is bid at a high enough price to cover marginal cost (e.g., cost to build the equipment, cost of installation and service), though not enough to make a profit when a vendor takes into account the cost of R&D, corporate overhead, and other indirect costs. These aggressive bids aren't sustainable—if every bid always just covered marginal cost, a company would slowly go bankrupt. But they help buy share in important markets or establish deployments that might showcase a new product. Call it a marketing investment.

Huawei's pricing with BT and others didn't look like a marketing investment. In bids throughout the world, as competing vendors submitted quotes within a few million dollars of each other, taking hits to their profit margins, Huawei was coming in with prices that didn't seem to make any sense. In some cases, the bids were so low they didn't appear to cover the out-of-pocket costs of third-party gear, like the heating and cooling equipment all companies have to buy from GE or other makers, let alone the vendor's network equipment. Lucent's Pat Russo remembers a competition her company participated in to build out a network in Southeast Asia. "We bid," she says. "Nortel bid. We later learned that we were about a million dollars over them, at $23 million. Then Huawei came in at $10 million." A bid like that didn't qualify as dumping—it couldn't have covered even the marginal cost on the project.

This may sound like sour grapes from a company that was getting its butt kicked around the world by the better, cheaper, harder-working upstart from China. After all, how could a company based in the United States compete with Chinese labor costs? But by then, Lucent was already taking advantage of the same low-cost labor in Asia, assembling many products in China, and incorporating many more low-cost components into the devices they assembled in other parts of the world. By the mid-2000s, virtually every network equipment maker was erasing China's cost advantage with their own China-based manufacturing.

These inexplicable quotations were widely confirmed throughout the industry, including by carriers with no ax to grind, who were all too happy to take advantage of the aggressive pricing. One executive, the former president of a major US carrier, recalls a wireless project he worked on with a non-US carrier. "The Huawei bid was so low," he explained, "that the other vendors said, 'We couldn't possibly match that. Our cost of the parts is higher than their bid.' No one could compete with them." That is, Huawei appeared to be selling with no profit margin, not even covering the costs of making, installing, and supporting the equipment. While they didn't win the business out of political concerns, the carrier issuing the contract did return to the other competing bidders and ask them to sharpen their pencils and submit more aggressive pricing. They complied.

So even when they lost, Huawei destroyed their competitors' ability to make money, reducing their profit margins and crushing their R&D budgets.

Huawei's pricing strategy effectively made their Western competitors' products worse. And it was brilliant in its simplicity. Huawei entered bids for every contract they could, everywhere in the world. They bid at prices they knew their competitors couldn't match, at least not profitably. When they won those bids, they denied their competitors the revenue from the win. But when they lost, the pressure to come in aggressively against Huawei, and the requests to lower their "best and final" pricing that the winners invariably received, wiped out their competitors' profit margins.

Huawei was destroying its competition simply by showing up.

There was just one problem with this bid-below-cost strategy: There was no way Huawei would be able to sell products for so much less than the rest of the world without eventually liquidating the company. The losses would have amounted to billions of dollars a year, enough to wipe them out in no time.

Yet, according to their reported financial results, Huawei was growing in all metrics: revenues, profits, R&D budgets, margins. The seemingly impossible pricing became Huawei's hallmark throughout their breakaway growth, with suspicions continuing about how they could bid below cost for decades and still report profits and pour billions into new investments. Years later, Samsung's executive vice president, Woojune Kim, responding to a question in a 2020 session before a British parliamentary committee, would testify, "We have frequently seen bids that do not seem to make sense in the pricing. No company beholden to shareholders and to make profits could offer that sort of bid." Kim went on to explain that Samsung, which is one of the few electronics manufacturers in the world larger than Huawei, was highly efficient in sourcing and assembling gear, but even his company couldn't approach a cost basis that would let them compete with Huawei, and that such pricing was "not sustainable."

Huawei's own government affairs team released a study they commissioned by a researcher named Dan Steinbock that compared anti-Huawei rhetoric to McCarthyism and boasted that after Huawei entered European markets, "profit margins plunged to 30–35%, which supported consumer welfare . . ." It was an odd combination of a boast and an admission. Reports from the European equipment suppliers support the claim that their margins had been destroyed by Huawei's aggressive pricing.

It can't be argued that Huawei was able to do this because, as a private company, they didn't have any shareholders and weren't worried about making profits. The need for profits isn't some capitalist construct that doesn't afflict private companies or businesses in communist countries. Profits for any company are the source of investment; they fund R&D, sales force expansion, training, and everything else that enables a company to grow. The only reason a company like Amazon was able to experience rapid growth for years without turning a profit is because they kept drawing on massive infusions of capital from investors who believed in their model and knew that someday they could ease back on the growth and let the profits roll in. By the time Amazon finally hit that point in 2001 and reported their first profit, they had consumed a staggering $2.8 billion in losses, funded by those investors.

Huawei may have burned through more than $75 billion.

THE DEEPEST POCKETS OF ALL

The *Wall Street Journal* thinks they figured out how. In a report published in December of 2019, a team of researchers at the *Journal* concluded that Huawei had received $75 billion in state support during its rise to world dominance, allowing the company to undercut rivals' pricing and still pour billions into R&D. The company received more than $3 billion in outright grants and land discounts, most of those since 2008, and enjoyed $25 billion in tax incentives. Huawei insists that they received only "small and non-material" grants, many of which were available to others. Founder and CEO Ren Zhengfei even told the BBC in an interview that his company received no government grants whatsoever, a claim his PR department was forced to walk back, acknowledging that, as their own annual report described, the company received significant preferential treatment and financial consideration.

The *Journal* found that Huawei had assembled the land for its massive new campus through uncontested auctions where they paid as little as 10 percent of the comparable market price, saving billions of dollars. The *Journal* also says that the mayor of Shenzhen acknowledged as much at a state conference in 2012, saying that "local officials began waiving or reducing levies on Huawei . . . in the early 1990s."

But don't all countries provide support to local industries, especially those deemed vital to national security and economic health? Aren't these arrangements comparable to those offered to Western "national champions"? Not within orders of magnitude. For example, from 2000 to 2018, Cisco received a total of $45 million in state and federal subsidies, loans, grants, and guarantees, according to the *Journal's* analysis—still less than 0.1 percent of the aid reportedly given to Huawei.

A December 2000 article by Bruce Gilley in the *Far Eastern Economic Review* reported that, in 1998 alone, the state-owned China Construction Bank lent Huawei nearly $500 million in buyer's credit. A lot for Huawei, but a staggering amount for the bank; it represented 45 percent of the bank's total credit for the year. Hardly a classical diversification strategy for a bank, and not one you might expect from an institution that wasn't even focused on telecom deals. But when the Party calls . . . Regardless, it was a drop in the bucket compared to what the Chinese government was prepared to provide to Huawei. *Wall Street Journal* research says that in the years after 2000, Huawei received more than $45 billion in loans and other support from many state-run banks, all under the control of the CCP.

These loans in particular would enable Huawei to create a machine that pushed their products out to telecom companies all over the developing world and win 3G, 4G, and 5G contracts across Europe and Asia.

WHY COMPLAIN?

Why didn't the regulators and government authorities in the United States and Europe stop the damage from Huawei? Why did they allow the Chinese company to sell their gear in the US, and all over the world, at prices that were clearly below cost, low enough to take business from most competitors, and so low they destroyed the margins on competitors who had to slash their own prices to win the business? Shouldn't countries prevent foreign companies from entering and dominating their markets, selling products below cost? And shouldn't countries put up trade barriers to block foreigners if their home country doesn't also open its markets to imports?

This is often the response from governments, whether in the Americas, Europe, or elsewhere. Leaders come to the aid of domestic industries, decrying unfair competition. Companies under threat try to position themselves as "strategic," and warranting protection. Somehow, the truly strategic telecom sector failed to earn this designation. But should any sector block foreign suppliers, even if those suppliers are selling their products at unreasonably low prices?

There aren't simple answers to these questions, says Mike Munger, an economist and former chair of the department of political science at Duke University, where he continues to teach political science, public policy, and economics. He argues that trade deficits aren't a problem, and he makes the case for continuing to trade even when a partner subsidizes their products and puts up barriers to yours.

"My trade deficit with Kroger's supermarkets is gigantic," he deadpans. "I buy a ton of groceries from them and they never buy anything from me. The last time I tried to sell them some of my stuff, they called the cops." Munger says that if one party has something of value and the other party has money they want to exchange for it, no voluntary exchange is bad.

The reason he gives the store his money is because Kroger can deliver food to him cheaper and better than he can produce it himself. In business schools, this is called the "make or buy" decision: never make anything yourself if you can buy it from another source for less. Munger instead chooses to focus on his core business, teaching classes and publishing books and papers. If he spends time making *other* things, things he could have purchased on the market, he's losing money. If he didn't already realize it, the market is telling him that society values his abilities as a professor more than his ability to churn butter or make his own shoes.

If Huawei sells us gear cheaper than we can make it ourselves, we ought to buy that gear and let our competing companies get into a different business, Munger argues. We would avoid expending labor creating something of one value and instead acquire that thing for less, freeing up our labor to create something *more* valuable.

"The division of labor is the only source of wealth the world has ever known," he says. "Wealth is the result of me specializing and you specializing: the total amount of stuff produced increases."

What if the company, or country, selling it funds those sales at prices below its own actual cost?

"Well, more power to them. That's terrific. If Kroger subsidized food to make it cheaper for me, I'd thank them." In effect, when the Chinese government funds Huawei's exports, that is a tax on its consumers and citizens and a transfer to America's consumers, who get stuff more cheaply.

And what if a company is just selling things cheaply so they can drive everyone else out of business and jack up the price? Munger explains that the math is clear that it doesn't make sense to do this. The up-front cost to sell so cheaply is typically too high for a company to earn it back through excess prices after they have become the last man standing. Other companies will just reenter the market if they see an attractive price environment, preventing the first company from reaping those excess profits. China subsidizing its exports to undercut our companies hurts its citizens and transfers the money to ours, and it can't expect the move to pay off, at least not economically.

Munger says that this process may not make economic sense for China, but that's *their* problem. Maybe it gives them an industry that they think they need. And China blocking our imports? Sure, it's better for us and them if they let us compete freely, but we would only make it *worse* if we created our own barriers too. Blocking or limiting American equipment vendors hurts China's companies and citizens, who get less choice and may have to pay more because of the restricted competition. But, again, that's not our problem.

Munger continues, "You might say, 'But China benefits from this more than we do!' That's fine. It means China's wealthier and it's likely to (eventually) start buying more US exports, because if it has more money, it won't be so export focused." In other words, wealthy countries consume more, and that means they should, eventually, import more. "If you look at South Korea, Japan, even Germany, who used to only focus on exports, they started buying iPhones and software and a lot of things that the US made once they became wealthy."

But then Munger pauses. "This is how free trade is good and creates wealth," he says, "but only provided that you take a liberal worldview that we're all in this together." This model of wealth creation only applies when you're dealing with trading counterparts, where the relationships are long term and the tone is primarily cooperative. As long as you get wealthier, you shouldn't really care if your counterpart gets *even* wealthier than you. It's only in a time of war that you refocus from *absolute* wealth and power to *relative* wealth and power vs. other countries.

Now, if Kroger's plan were to undercut all other supermarkets until they went bust, then suddenly cut off the food supply and starve everyone into submission, that wouldn't be a very good business plan. It wouldn't make much *economic* sense to harm their customers, and they would lose a fortune subsidizing the food long enough to destroy their competitors—more money than they could ever make back price-gouging later—but they would certainly gain power over their customers, far more than a grocer might typically have.

And that would be an entirely different story . . .

The End—For Now

For China, purely economic success is not an end in itself.
It is a means to wider political and strategic objectives.
—William Barr, former US Attorney General

By 2006, the downturn in carrier spending and the price pressure brought on by Huawei's aggressive bidding had pushed Lucent to the brink, to the point where the General Services Administration (GSA)—the government agency responsible for all federal purchasing—had reached out and expressed concern about Lucent. Was this an offer from the government to help pull them through the crisis? Not quite.

"Because we were in financial difficulties," says CEO Pat Russo, "the US government decided they weren't going to buy from us!" She was stunned by the news. "I'm fighting to bring the company back from the brink of bankruptcy and I have to try to convince my own government not to terminate our contracts! They should have said, 'This is the organization that has Bell Labs, this is the company that created all these wonderful technologies and capabilities. We ought to be supporting them, not making it even harder for them to do business!'"

For those who oppose corporate welfare, there is a difference between bailing out a bank that has screwed up and ensuring the ongoing existence of a company that helps provide national security. The government could have allowed the shareholders to be wiped out—that's the penalty to the company's owners for the financial situation—and then reconstituted the company in a number of ways, either finding a US buyer or recapitalizing it with money from the private or public sector. The federal government, which is so assiduous about protecting soybean production and textiles from foreign competition, decided not to intervene with the impending failure of one of the country's leading research and development institutions.

Lucent, having stared down the abyss five years earlier, finally hit the end of the road. The company entered talks with European vendors, including Nokia, but Alcatel emerged as the rescuer when they offered $13.5 billion to acquire the entire company. It was a far slide from Lucent's market value of $258 billion just a few years earlier, but a premium to the current trading price, and a path out of collapse. The offer was for $3.01 a share, 96 percent less than the company's previous peak of $84.

Lucent took it.

This was the end of the equipment legacy of Alexander Graham Bell, the company that built the pay phones that were indestructible to every insult but progress, the standard-bearer that supplied the country's wired, wireless, and internet service providers with the most advanced, gold-plated network equipment in the world.

And it was the end of Bell Labs, inventor not just of cellular telephony, fiber optics, the *transistor*, but also the company responsible for innumerable "black budget" operations for the US government. At least, the end of them as an American company. And this presented a problem. The flagship of technology innovation was about to find itself a part of a company housed in a foreign country, and it would need to find a way to partition its most sensitive activities from foreign eyes and influence. It mattered little that Alcatel was a publicly traded company with full transparency of operations, based in a NATO country that had been a steadfast ally of the United States since . . . well, since before the US was a country. Now, at last, the US government came to the realization that it might be a problem that a foreign country would have its hands on the equipment running our networks, that the French could be trusted to safeguard the intellectual property and current development programs of Bell Labs.

The Lucent board was similarly concerned and retained one of the biggest consulting and systems integration firms to develop a plan to relocate and secure Bell Labs' sensitive operations and intellectual property. The consultant on the project, Mike Johnson, took responsibility for implementing and supporting controls with the Committee on Foreign Investment in the United States (CFIUS), a federal interagency committee that consists of nine cabinet members, including the secretaries of state, defense, homeland security, and commerce, as well as various other agency heads appointed by the president. Their role is to make sure that the acquisition of an American company by a foreign-controlled entity will not compromise national security. (There is no organization or review process that exists to make sure a foreign-controlled entity doesn't achieve this by driving an American company out of business.)

Johnson's first step was to propose the structure for a subsidiary, Lucent Government Solutions, which would house the programs at Bell Labs that were classified or had a material impact on national security, like those supporting the NSA or CIA and anything involving classified weapons systems. "After we put all the cool stuff into LGS, we got an unexpected call," recounts Johnson. His team was gathered in its own project war room, reviewing the work plan when the phone rang.

A woman's voice on the other end said, "Mr. Johnson? The general would like to see you here at his office at the Pentagon."

"The general?" Johnson covered the mouthpiece and turned to his partners on the project. "How did they even get our names?"

But Johnson and his team didn't argue. Within a few days, the senior leadership from the company's account team arrived in Arlington, Virginia, and after clearing the security checkpoints at the Pentagon, they joined the general and his staff around a conference room table. The general spoke first.

"So, you're the guys who are handling the spin-off of Lucent Government Solutions. How'd we do? Did my guys ask the right questions? Are you confident we have the right security measures in place?"

Johnson looked sideways at his colleagues before replying, "Uh, no. Well, yes and no. Your guys didn't know what they were doing, not any of them, but yes, we're taking care of the security issues, and we're confident we'll have adequate measures in place."

"*Really?*" asked the general, who hadn't missed the sideways glance. He leaned forward, hands now on the table. "What are you saying?"

"Don't worry," Johnson assured him. "The company is doing the right things anyway." He explained that they were splitting Lucent apart and securing elements that presented risk of compromise from a foreign country, including classified lab facilities that couldn't be physically relocated from the soon-to-be French-owned Bell Labs buildings. He continued, "Your guys showed up looking for door locks, asking us about physical barriers between adjacent work areas. But they didn't ask about the important stuff, like securing the WAN."

"The WAN?"

"The wide area network. The US telecommunications network. It runs on Lucent's 5ESS switches. They underpin all communications in the country—private, public, corporate, governmental. Even military. And your guys didn't ask us to make sure the French couldn't put code in there. Like putting a software update on the 5ESS that added a back door."

The general's jaw tightened. He looked at his staff sitting around him as they fidgeted in their seats, considering, perhaps for the first time, what the loss of America's primary telecom equipment maker truly meant. In the hands of another company or another country, the switches running American networks could be remotely modified to allow intercepts between parties, or even throttled to shut the network remotely. Sure, French scientists wouldn't be able to walk into a secure part of the lab that was working on a classified project for the Department of Energy, but perhaps they could intercept conversations on America's phone network. Including the communications between offices of the Department of Energy. Or if those conversations were encrypted, they could still know who was talking to whom, and when. It's in the metadata: Even when they couldn't hear the content of an encrypted call, whoever managed the network knew who called whom, when, for how long. They knew what websites they visited. They could see where and when traffic was building. They could also disrupt or terminate the ability of parties to complete calls on our own network.

It wouldn't even require a hack. The supplier of the network equipment—once Lucent and now Alcatel—was required to leave its hands on that deployed equipment, updating it with new software, adding features, patching security vulnerabilities. And those updates didn't pass through CFIUS or any other review board. Not that it would matter. The complexity of the code—millions of lines written mostly in C language—was beyond anyone's comprehension apart from the people running that equipment. Who would now be based in a foreign country.

The Pentagon officers, expert at what they did, were in way over their heads when it came to telecom equipment.

"They were trained in guns and bombs," says Johnson. "They didn't understand what this really meant to national security."

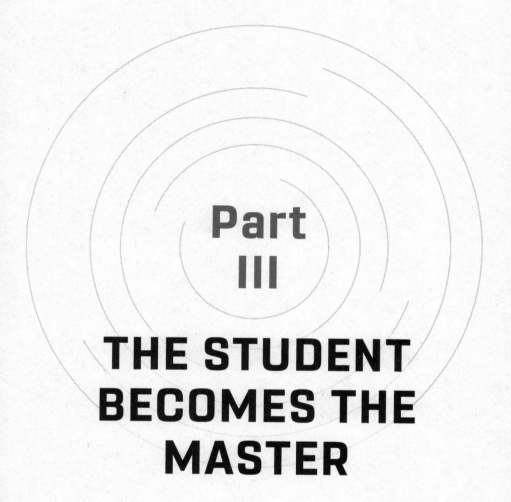

Part III

THE STUDENT BECOMES THE MASTER

15

Culture and Credit Cards

He who does not trust enough will not be trusted.
—Lao Tzu, Chinese philosopher

As the former industry giants reeled from their losses and consolidated to cut costs, Huawei was on the offensive, snapping up staff and moving to grow share in the developed markets that had eluded them, above all the United States. There were cultural barriers that they would have to clear, many of which they were thoroughly unprepared for.

Hiring new staff was easy enough. In 2008 the United States was entering an economic downturn, and the telecom equipment sector had still not recovered from the crash years earlier. Huawei was able to pick from the cream of the American telecom equipment industry, luring people away with higher salaries, snagging them after layoffs, cherry-picking the best with guidance from former colleagues who had already joined Huawei. Some were engineers and technicians. Others were field support staff, brought in to deliver and install the new Huawei cellular base stations. Many were sales people, hired to deliver the *guanxi*, the local relationships, that money couldn't buy.

Of course, there were some relationships money *could* buy. For the right price, a sales rep with a long-standing, trusted customer relationship would drop off his badge, turn over his PC, and sign on with the new, higher bidder. And if you could lure away a salesman who had sold hundreds of millions of dollars of Nortel gear to Sprint by paying him 30 percent more, you were starting out with a personal seller-buyer relationship that ran deeper than any corporate brand. Even for the reps who had survived the cutbacks—the top salesmen who were secure in the remaining jobs at Alcatel-Lucent, Nortel, Cisco—the lure of a better pay package always warranted a return phone call. Sales reps are coin operated; you pay, they sell. And if you're willing to pay more, most are willing to take their book and carry it across the street.

But what, exactly, would they find there? It wasn't easy to learn about Huawei; compared to a Western company, their website told very little. There were pages about the products, the services, the promise of support and upgrades and continuing R&D investment. There were beautiful pictures of the gear and the gleaming new headquarters somewhere in China. There were exhortations about the glory of the company and its workers—comical, almost, for a Western tech worker who didn't understand the Asian philosophy of employee-employer relationships and the schmaltzy rhetoric that characterized their internal corporate communications. But there was not much about the people running the company. In fact, there was nothing on the management team of Huawei. There was no web page with pictures of the senior leadership, giving their bios, their responsibilities. Not even their names. No board of directors, no CFO, no data on the company's governance. No organization chart. If the company seemed cryptic, these new hires were about to learn one thing about how Huawei was managing its efforts to grow in the US. Which is to say, not very well.

BUNKING UP

One consultant, brought in to help Huawei figure out its personnel issues, describes a scene that unfolded in an overcrowded conference room in Huawei's Plano, Texas, office, where the latest hires were getting their orientation. Some of the people had been working there for a few weeks and were finally taking in the on-boarding lecture. Others had signed on in the prior days and were just getting their first few minutes of what it was going to be like working at this Chinese upstart. All of them sat in aluminum bridge chairs, notebooks on their laps, faces turned toward the human resources manager as she welcomed them to their new jobs in good, if not perfect, English.

"Please do not incur expenses and do not submit anything to me for purchase until the beginning of next month," she began. For the employees who had been on the job for a few weeks, these words seemed odd. They *had* to incur travel expenses in order to see customers, attend conferences, buy office supplies. And for some of the people shifting in their chairs, these words sounded ominously familiar to what they had heard as the wheels came off their former employers. Billion-dollar companies had suddenly run out of cash. Contracts were broken, bills left unpaid, newly built office complexes never unoccupied. But that wasn't the case here, with Huawei. Was it? Had they been misled? Would this turn into one of those stories of gross incompetence where a company had hired thousands of new employees only to turn around weeks later and cut them all loose?

Not exactly.

"My Amex card is at its maximum," the HR manager continued. "We have too many people incurring expenses at once, too many purchases for one card. Please wait until this statement is paid and I can start taking your charges again."

One new hire sitting near the door—a well-known sales VP recruited from one of the largest equipment makers in the world—grimaced. "You gotta be kidding me," he said. "They're using one Amex card for the whole company's corporate travel, purchases, dining, entertainment? Do they not trust their own sales executives with their own corporate credit card, or are they just way behind in setting up the most basic corporate processes?"

It was both, actually. They hadn't set up an operation yet to cope with the rapid increase in the number of new employees. HR systems, expense management, training, all the normal corporate functions hadn't been geared up to handle the newly constituted American division, and business operations were way behind where the hiring team was.

Quite apart from that, there *was* no trust in these new workers. It wasn't the Chinese business culture. Field technicians in China weren't issued corporate credit cards and told to "be good." Even the sales reps in Shenzhen didn't work that way. The controls were tighter; the faith didn't run so deep.

More troubling to the new hires, there seemed to be no understanding of the cultural universe in which American companies operated, and in which these employees expected to work. Beads of sweat were starting to break out on the foreheads of former service technicians who had just picked up their new business cards.

The orientation continued. "When you travel to call on a customer, please think of who you know in that city or town. If you have family, friends who live there, perhaps they have a guest room or a couch that you can sleep on. This isn't a vacation—you should do everything you can to keep the costs down and not worry about a fancy place to stay."

As the consultant observing this now recounts, "You could feel the tension build in the room. Were they really asking their top salespeople to sleep on friends' couches during business trips?" But the best—or worst—was yet to come.

"If you do not know someone in the town who can provide accommodation," the HR manager continued, "we understand you may need to stay in a hotel. Please pick a reasonable location, and if you are making this trip with a colleague, you are required to share the hotel room with them." She smiled. "We will provide a fifteen dollar a day reward for this."

The sound of a notebook snapping shut startled the new hires out of their bewilderment. Some looked up in time to see the sales VP, less than an hour into his new job with Huawei, shake his head as he stood up, pushed the escape bar on the door, and walked into the sunlight of the building lobby. He stopped at the front desk and dropped off his laptop, with the protective plastic still covering the Lenovo logo, ignoring the confused look on the receptionist's face, and headed out to the parking lot.

Huawei was the new game in town, but that didn't mean they understood the rules.

YOU COULD HAVE JUST ASKED

Even as Huawei was learning how to hire in the United States, they were bringing American consultants to their massive headquarters in Shenzhen. They weren't looking for help in working their way into the US market, but they were seeking to take advantage of the more creative and strategic thinking delivered by America's technology consultants. The company was hardly in trouble; the help they needed was to manage explosive growth and ensure they were positioned to continue it. Revenue for 2008 had more than doubled to $18 billion from $8.5 billion in 2006, and margins and profits were growing rapidly too. For an equipment company of this size, such growth was unprecedented.

Even if American consulting companies had reservations about taking on a project working for a Chinese tech firm, the lure of a rapidly growing high-technology customer was irresistible, especially after the financial crisis led to a collapse of much of their customer base. The big US consulting firms rushed to fill the need. A partner at one of them, Tom Miller, found himself climbing aboard a flight from Dulles Airport in Virginia to San Francisco, and on to Hong Kong before continuing to Shanghai, preparing for what might be a long assignment. In fact, it would be six months before he finished the project, having not just educated his clients on global growth strategies but having received an education himself on the cultural differences between the way the United States and China did business.

Miller's team was brought in to help Huawei plan the rollout of a complete set of new services, defining the operating model and the market strategy. As aggressive as the plan was, one thing was missing: There would be no rollout in the US. Huawei was still effectively excluded from any American deployments, at least any large ones, so the focus was on sales to Africa, Latin America, and southern Asia.

One of Huawei's executives explained their position to Miller over dinner on his first night in town.

"We're not industrial grade like Ericsson is," he acknowledged. "At least, the bigger carriers don't think we are. Ericsson just has better stuff right now." This wasn't an admission of inferiority. It was just a recognition of the patience Huawei would need if they were to eventually become a player in the developed countries' markets.

"They were going to grab share in their current markets, trial their new systems, and improve them," Miller explains. "And they were going to undercut everyone on price across the southern half of the world, essentially. After their aggressive pricing weakened the more established incumbents like Nokia and Ericsson, after they'd won all the deals in the developing countries, then they would close in on Europe and North America."

Over the years, Miller would occasionally give presentations on his experiences working with Chinese tech firms that were pressed by the CCP to establish technology leadership. Miller remembers his Chinese clients as being bright and exceedingly hardworking but lacking in some of the business thinking that characterized their counterparts in the US and Europe. "What I realized was that they could memorize every detail of a complex chart," he says, "but they didn't necessarily understand how it all flowed together." They also seemed to be working on a different business ethic, as he discovered early on, and in a peculiar way: "They hacked my hard drive in my first week there."

He recalls giving a presentation to the Huawei client team, showing them slides he had prepared the night before, when he was asked an odd question.

"Mr. Miller, could you please explain the chart on slide eighteen showing the incumbents' share of the India market?"

"Well, yes . . ." he replied. "I was, uh, *getting* to that . . ." Flipping ahead a few pages, he pulled up a slide he had just created on the topic. It became clear that not only did the client team already know what was on his laptop, they weren't embarrassed to let that slip. Just get on with it. His team took note of that in how they handled future work and even started using cheap "burner phones" they picked up in Hong Kong and discarded weekly.

It was hard to wrap his mind around it, but Miller learned that this was simply their business culture: "Several people on my team had been born and raised in China before coming to the US for college. They saw my puzzlement and explained the difference to me. They told me that they're not just stealing *our* stuff. They're stealing *each other*'s stuff too. That this sector, as least right now, is working under a totally different ethical framework. They take care of the ones they feel they're responsible for—company, family, country. If they're working for Huawei, and Huawei is helping them succeed, of *course* they would steal ZTE's stuff, of *course* they would steal Cisco's stuff. ZTE's trying to steal your stuff; you go steal theirs. It would be *wrong* not to seize an opportunity to support the party you're responsible for. It's a free-for-all. You're in one circle or another; all your calculations when you're interfacing with other firms come down to 'Are we in each other's circle? Are we working to steal stuff from other people?'"

Some people try to understand this cultural clash by looking to the difference between Confucian philosophy and Western ethics systems based on religious ideals. Such analyses have value, but the subject is a complex one, fraught with the dangers of painting an entire people with too broad a brush. American business ethics isn't defined by the behavior of 1980s Wall Street investment bankers any more than by the values of a shopkeeper in Vermont. But by most accounts, anecdotal and in the legal record, the aggressive behavior by China's telecom companies during the years around the millennium ran counter to most of the world's business nature, whether in Korea, Europe, the United States, or elsewhere. This lack of alignment between business cultural values, at least as claimed by Western observers in the telecom business, was bound to lead to conflict.

Brain Drain

Trust but verify.
—Ronald Reagan

Huawei wasn't only hiring salespeople and consultants. They also started tapping highly specialized technical talent, engineers, and scientists from Bell Labs and especially Nortel. Anyone with wireless expertise.

Most people doing a technical job don't worry about the politics of their employer; working for a foreign company, political policies of the hiring company's country are an abstraction, not relevant to engineering calculations that transcend language and culture. What matters is the job, and as with all jobs, the experiences of people recruited to join Huawei varied. Some got rich serving them. Some got bored. A few got fired.

And some quit in disgust.

Ron Marino didn't think twice about turning in his badge and leaving Huawei. It wasn't about the money, that's for sure. The company was paying him 30 percent more than he had been making at his previous job, and the money had been good there too.

It wasn't about the working conditions, either. They were tough—exhausting at times—but he was used to working on massive projects with impossible deadlines under enormous pressure. His decision to leave was more a visceral reaction to the way the company seemed to view him. What they wanted from him.

CROSSOVER SKILLS

Marino had worked for some of the biggest and best names in wireless technology. He had designed radio systems for Litton Industries, a defense contractor that, while not as well known as the giants like Lockheed, led in critical areas of electronic warfare systems and microwave navigation and communication. Litton's radio technology was an integral part of the Pentagon's tactical and nuclear military capability, and Marino was proud to have been a member of their team.

As a requirement of his work with Litton and, after the company was acquired, with Northrup, Marino had to pass extensive security clearances and sign nondisclosure agreements. Marino took the responsibilities seriously and went beyond the legal restrictions on sharing information later in his career, refusing to discuss what he considered sensitive information with colleagues who weren't "cleared." He had never felt pressed to divulge confidential information in most of his later civilian projects—but, as always, Huawei was different.

Marino was recruited to the job in 2011 by his former boss, another American who had been a top radio scientist at Bell Labs, spending twenty-five years at the company and achieving the ultra-elite distinction of "Bell Labs Fellow." When Marino got the call about a new job opportunity, he was all too happy to join. A dozen of his former Bell Labs colleagues were coming on board, too, a wealth of talent and know-how.

The company hiring them, Futurewei, billed itself as a US-based entity, a research affiliate under the umbrella of Huawei Technologies Co., Ltd. If the sales reps hired into Huawei's Plano, Texas, operations had trouble learning about the company, it was no easier for Marino. He couldn't learn much about Futurewei from their website—it told him nothing. The company was a thinly veiled front for Huawei, though so thinly that they had no identity, no logo, no branding at all that was distinct from the parent company. They were Huawei in America.

Futurewei's interview process was conducted entirely by Huawei engineers, and the process was different from anything Marino had been through before. Typically, these types of interviews consisted of a screening to see if the prospect really knew how to do the kinds of things he would be expected to do. Did he demonstrate competence, was he confident in his technical knowledge, was he smart enough to deliver as expected? But this interview was focused on where he had worked, which projects he had knowledge of, and what information he had of existing Western antenna technology.

"They wanted me to bring slides of everything I had worked on," says Marino. "I was damned if I was going to hand this stuff over."

NUCLEAR
ANNIHILATION

Just a few years prior, Marino's experience with antennas wouldn't have been of much interest to a company looking to secure a competitive advantage through cutting-edge technology. Antennas in older wireless networks were pretty low tech: cheap, simple boxes that beamed a signal out to whatever cell phones they were pointed at and picked up their return message, dumb units that passed the signal down a coaxial cable to the radios for processing and forwarding into the network. For forty years, since the first cellular systems were deployed, antennas were fixed units, bolted with U clamps to the tops of towers or screwed into the sides of buildings, pointed toward the area needing coverage. That's what the antennas *used* to be, before the advancements were proposed for new 5G networks to improve speed and network capacity.

But these advancements weren't actually new at all. The transformation of cellular antennas from dumb boxes to advanced technology predated the development of 5G or 4G wireless networks. In fact, it predated cellular telephony altogether by nearly twenty years.

This technology was ignited following a meeting in the Kremlin, when Soviet leader Leonid Brezhnev and American president Richard Nixon sat down to discuss how to ensure that their respective countries would be completely wiped out in a nuclear war.

In the decades after World War II, the two foes amassed massive nuclear arsenals, with both sides spending hundreds of billions of dollars developing, deploying, and maintaining ever-more sophisticated weapons systems. As the intercontinental ballistic missiles grew in power from kilotons to megatons, both sides were drawn to the negotiating table to find a way to cut their expenses without giving up a military advantage. One such proposal pursued a macabre policy called "Mutual Assured Destruction," or, quite appropriately, MAD. The underlying concept was that neither country would ever launch a nuclear missile attack on the other if it was certain that the response would annihilate the aggressor's country too. That is, each side had to believe that, even if it launched a successful attack against the other, the opposing side would retain enough offensive capacity to launch a retaliatory strike that would wipe out the aggressor, and probably the rest of the world in the process.

But this could only work if it were certain that all or most of the country's launched missiles could hit their targets. And that meant that both countries had to agree not to develop or deploy systems that might stop the incoming ballistic missiles; for this model to work, neither side could contemplate the survivability of a nuclear exchange.

The result of the negotiations was the Anti-Ballistic Missile Treaty of 1972, banning almost all antimissile systems. In that agreement, signed by President Nixon and Chairman Brezhnev at a Moscow summit, the countries promised to leave themselves vulnerable to incoming slaughter from their mortal enemies. The savings to each country would come by eliminating the need to build even larger arsenals in order to get through a missile defense system.

Some may have believed that the Soviets intended to honor this suicide pact, but Ronald Reagan wasn't one of them. When he was elected eight years after the signing of the treaty, he commenced his "trust but verify" approach to the Russians, which was heavier on the "verify" part. And verification efforts soon showed that the Russians were violating the ban by deploying antimissile radar systems beyond the agreed terms. Soon thereafter, Reagan set about developing a system to protect America from incoming missiles, whether launched by the Soviet Union or any of the handful of other nuclear powers or rogue states. His Strategic Defense Initiative, derisively referred to as "Star Wars" by his political opposition, called for a solution that could shoot down incoming missiles using advanced radar and tracking technology—technology that not only didn't yet exist but was considered impossible by many of the leading scientists of the day. Radar that had been around since World War II was capable of identifying an airplane traveling at hundreds of miles per hour, at altitudes up to 80,000 feet, about fifteen miles high. ICBMs weren't such an easy target. At the peak of its trajectory, an ICBM may be more than one hundred miles above the earth's surface, traveling at more than 15,000 miles per hour. Tracking such an object with enough accuracy to shoot it down was a challenge that pushed the limits of technology, not least of all radio and antenna technology.

The solution, as developed by top scientists at a small group of American private and government facilities, including Bell Labs, was initiated during World War II and relied on something called "phased-array antennas." The idea was that by setting up a coordinated array of lots of small antennas, controlled by computers, a radar device could track small, fast-moving objects far better than using traditional fixed antennas, effectively put more energy into the signal bouncing off that object with less energy wasted pointing at the background and empty sky. By combining this technology with other "smart antenna" software and hardware, scientists were able to develop far more effective tracking systems. And with new "beam forming" capabilities, the radar could follow the object as it moved across the sky at high speeds.

The extraordinary demands of tracking a missile or fighter jet led to arrays that sometimes required hundreds of closely packed antennas. Fitting so many into one box required using tiny antennas, which meant operating in high frequencies at very short wavelengths. (The size of any antenna—your cell phone, car radio, or a military defense system—is determined by the size of the wavelength it's sending and receiving.) In order to pack that many antennas into a compact area, the military would use radios operating in a part of the spectrum that had extremely high frequencies, like 20 GHz, 66 GHz, or higher, unlike the 2 GHz or 3 GHz typically used for cell phones.

This high-frequency section of the airwaves—a backwater of little interest outside of military applications—would suddenly become interesting to the wireless carriers around the world when they began looking for ways to deliver more bandwidth than they could fit into the crowded spectrum used by 3G and 4G systems. This high-frequency part of the spectrum, called "millimeter wave" because of the short wavelength, was embraced by some carriers like Verizon because of the ample bandwidth available in that part of the spectrum, despite its difficulty in penetrating objects in its path. For a military radar pointed at the sky, this isn't such a problem.

As the military phased-array antenna technology became more robust, affordable, and reliable, it trickled down to the commercial world of cellular communications, where 5G developers began to seize on the market benefits. Engineers at the telecom equipment makers realized that, by directing a cell tower's signal where it was needed most, the cell site could provide better service at a lower cost and with less power consumption. This might mean that a cell tower connecting to cell phones would aim its radio beam at a highway during the morning rush hour, lift it to point at office buildings during the work-day, and shift it over to cover a residential area at night, all without physically moving the mounted antennas. Even better, the antenna could identify a person who was trying to stream an episode of *30 Rock* on her iPhone and direct a signal straight into the phone, instead of wasting it on nearby classmates' phones that were idle or putting a lower demand on the network. It could even follow the individual as she drove to school, tracking the phone as it moved down the high-way—no mean feat, but easier than tracking an ICBM. All of this tech-nology was originally developed—often in violation of international treaties—to keep Americans safe from Soviet annihilation.

The result was that, by the twenty-first century, sophisticated and highly classified military technology had been repurposed to work in commercial networks, performing the same functions but with a differ-ent end result, and operating in a different part of the radio spectrum.

It was hardly lost on the Chinese that the commercial 5G tech-nology had a lot in common with the classified technology that could be used to shoot down a fighter jet or missile. And that the engineers working for cellular antenna companies might know a thing or two about defense systems.

THEY WANT EVERYTHING

Early on, Marino was asked to travel to China to "debrief the team," the first of several such trips, and an unusual request for an American work-ing in America for a supposedly American division of a company. The email Marino received just prior to departing from Newark International Airport gave the first hint of what Huawei wanted from him: everything. It was a prep sheet for his meetings with the people at HQ in Shenzhen, and an odd one. It asked him to come prepared to answer questions on "the following topics" and then listed *every* topic that could be remotely considered a part of base station or antenna technology. This didn't give him confidence in their understanding of what he did for a living, or, really, how the sector worked at all. But they knew exactly what they were doing.

Radio technology may seem monolithic to an outsider, but it is a highly fragmented field made up of ultraspecialized areas of incredible complexity. So much so that, in the nineties, when Bell Labs named a new head of forward-looking wireless research—a man with a PhD in electrical engineering from Cambridge, with a focus in radio technology—he was considered a "non-technical department head" by some of the staff because his PhD was in the "wrong area" of radio technology.

The emailed questions Marino received didn't seem to reflect this understanding. They asked him *everything* about antennas. Design. And operations. They had lists of questions on hardware, firmware, software. Electrical. Mechanical. It was like asking a medical doctor to come to a meeting prepared to lecture peers about hip replacement, heart transplant, acne treatment, psychiatry, and infectious diseases.

Marino knew it was unreasonable but figured that maybe it was a generic question sheet and he would only be expected to speak on his own area of expertise. He sat down with the questionnaire and started putting together some notes on his Lenovo laptop. The flight was in a few hours, and he had a lot of work to do.

Marino had never been to China before, and when he emerged in Hong Kong from the eighteen-hour flight, he felt lost. He didn't speak a work of Chinese and had little guidance on how to clear customs. One thing he did have was the phone they had given him in New Jersey, which he was to use to call his driver when he landed.

The phone gave him comfort, until he collected his suitcase from the carousel, hit the power button, and saw the "No SIM card" message on the screen. They had failed to activate the chip in his phone, effectively turning it into what the industry calls a "brick." Marino was off to a rough start. But he did have his driver's phone number and, after asking a passerby to borrow his phone, he was able to contact the driver and arrange a meeting spot. An hour later, he was in the hotel, and soon after, walking onto the Huawei's Shenzhen campus.

WE ARE . . . HUA WEI!

The place was massive—offices, factories, cafeterias, dormitories for thousands of workers. Even for someone who had received his degree in electrical engineering from Penn State, with 46,000 students at the main campus, Marino was dumbfounded, especially when he saw the exhibition center. "It was like walking into the Smithsonian Air and Space Museum," he says. "Massive. High ceilings. And they had every piece of technology related to wireless communications."

As Marino was brought into the research center, he flashed his new Huawei pass but was stopped by the guard posted at the door, who said something to him in Chinese and held out his hand. Marino looked at his guide. "Your phone," he explained. "Please turn over your phone before you enter the building."

Marino had worked at facilities carrying out classified government work, and he was used to strict security procedures in secure areas, but he had never seen a company that took cell phones from its own employees at the door. He handed his over. It didn't work anyway.

As soon as he entered the work areas he was struck by an alarming sight: At every desk and workstation was a large gun safe. He spent the next eight hours puzzling over this. Were the workers armed? What were they ready for?

His guide didn't offer any explanations. Marino was led past this apparent arsenal and into a conference room where he was met with his own fusillade, this one directed by an interrogation team of a dozen Huawei engineers. Actually, a dozen at a time, because that was all they could fit into the room. They were constantly rotating the people who came in to question him, but they all had one thing in common: they wanted to know about his previous work.

"All day long, the technical team peppered me with questions on antenna technology and radio design," Marino recalls. "'How can you get around this patent? How can we do this?' For five days straight, my mind was spinning."

Marino remembers getting uncomfortable at one point, during his first day of meetings. "They were touching on questions about what I call 'poor man's phased-array technology,'" he says. "The black magic isn't patented;

it's better just to keep it a secret, rather than publish your knowledge and have to rely on other countries' respect for intellectual property rights to keep it from misuse. I wouldn't divulge what I considered to be sensitive information, even if it wasn't classified. I'm reluctant to export my 'Yankee know-how' to anybody." And one other thing bothered Marino.

"Change the frequency and this is military stuff."

For Marino, six months in China felt like six too many. He turned in his badge and went back to work for a US-based company.

ZERO TRUST

Before he left, Marino did discover the secret of what was in the safes.

They were empty.

At least, during the day. But as he headed for the door to pick up his cell phone after his interrogation on that first day, he noticed the workers were clearing their desks and piling all the materials into the safes. Their research notes, their technical projects, whatever they had been working on was being locked behind a quarter inch of hardened steel.

"Nothing was left on the desks," Marino says. "*Nothing.*"

Who were they keeping it away from? It wasn't likely that Huawei feared infiltration from Nortel employees breaking into their labs in the middle of the night. Huawei had hired all the good ones anyway, after the collapse. There was a threat much closer to Huawei headquarters that gave them concern.

"You had ZTE down the road," Marino explains. Their home-country rival was a fierce competitor, constantly battling them for market share. They wouldn't think twice about taking intellectual property from a fellow Chinese company.

But there was *another* threat the workers also worried about, inside their own facility.

The guy down the hall.

The incentive systems at Huawei, and the culture there, led to fierce internal rivalries. Ren Zhengfei didn't only pit his executives against each other—remember wolf culture?—their divisions were in constant competition, and it wasn't unheard of for a group to learn of another team's breakthrough, steal it, and claim it as their own.

As consultant Tom Miller observed, their motivation and loyalty were wherever they felt their responsibility lay. Sure, they were all on the same China team, competing against the American, Korean, and European vendors. But within that China team, the competition was fierce against ZTE and Xiaomi, the world leader in low-end smartphones. And even within Huawei, each worker had a higher loyalty to his own division, and then to the work group within it. These engineers were making sure that no fellow Huawei employee would be able to steal their ideas and win a promotion or budget increase.

As Huawei worked to extend the company's share in other markets, they were confronted by values that didn't always align with their own. In the United States, these values would clash as Huawei undertook an expensive and ambitious effort to win over the authorities that were keeping them out of the market.

Hunting Big Game

The only difference between men and
boys is the price of their toys.
—Malcolm Forbes, American entrepreneur

If Huawei was having trouble understanding how to connect with American engineers, consultants, and sales reps, they had a better handle on how to engage those at the top of the food chain. In fact, they seemed remarkably adept at dealing with wealthy and powerful people; if salesmen were coin operated, the senior guys just required larger denominations.

These powerful contacts would be needed. Low prices and solid products were starting to crack open markets in the developing world and across parts of Europe, but Huawei realized that they needed more than that to crack the United States. They set out to acquire the *guanxi* needed to address national resistance to the company, as well as to strengthen relationships with the decision makers at the big carriers.

WHO'S THE BROSS?

In 2009, Huawei hired a new person to serve as their global chief technology officer and run their development worldwide: Matt Bross—the same Matt Bross who'd just finished overseeing the deployment of the British Telecom contract he had awarded to Huawei. While his new job was ostensibly to oversee the company's R&D, this description was met with skeptical responses around the industry. His style was freewheeling, out of the box, an iconoclast brought into companies like BT to shake things up and snap a company out of its inertia. *Mad Matt.* With Huawei's sales soaring around the world, they weren't looking for someone to knock the company off track through a radical new approach to business. So what was the deal with bringing on someone like Bross?

Controversy had dogged Bross since his days running service provider Williams Telecom's technology group, where he had been accused by some of participating too closely in the financial efforts of companies who were selling equipment to Williams during the boom years—this was a time when it wasn't necessarily illegal for a customer to accept pre-IPO stock in a vendor. A single contract with a service provider like Williams could make or break a start-up company, and the restrictions were lax about who could invest in a pre-IPO company. But the formal accusations that had been leveled against him were tossed out by a federal judge, and Bross's reputation as a gregarious, imaginative scientist with relationships throughout the industry made him interesting to Huawei, who needed someone to work with US leaders.

In fact, Huawei brought Bross on to land one American account. A huge one. And from the location of his new office, there was no need to guess which company was his target.

Bross would be leading a Chinese company's global technology development from an office in Wentzville, a small town in rural Missouri. A *home* office. In a post-COVID world, the idea of a senior executive working from a home office is not unusual, but at the time such an arrangement was nearly unheard of. Bross had grown up near there and was all too happy to return after his travels around the world on behalf of British Telecom, and his home had the advantage that it was near the headquarters of Sprint. Not close enough to pop over for lunch, but close enough that driving to Sprint's headquarters was faster than flying.

It was clear that his real job was to help win the Sprint account and, along the way, augment and build the Huawei technical team in the US.

LANDING SPRINT

At the time of Bross's hiring, Sprint, the struggling number three mobile service provider, was sinking in most metrics—stock price, subscribers, quality of service—but they rose to the top of Huawei's prospect list for a good reason. Two years earlier, in 2007, the wireless communications sector in the United States had consolidated extensively through mergers, acquisitions, and failures of smaller players, to the point that there were few large service providers left to buy gear. Verizon and AT&T were two of the titans, each with more subscribers than all the also-rans combined. T-Mobile was struggling to break through the 8 percent market-share barrier, and their future was uncertain. But Sprint was a contender, with nearly 17 percent market share and a strong motivation to change the game.

Sprint had been losing so many customers and suffering from such poor service delivery that their existence as a going concern was coming into question. On February 28, 2008, the company stunned Wall Street when it reported a loss of $29.3 billion—including a $29 billion write-down of Nextel—canceled its dividend, and took on $2.5 billion in new debt. Sprint's stock plunged nearly 10 percent that day, and analysts started expressing doubts that a turnaround was possible.

But not everyone took this news with such pessimism. A company this big—and this desperate—was a perfect prospect for a vendor trying to break into the US market with a promise of rock-bottom prices and swarming customer support. Huawei figured they were the perfect supplier to solve Sprint's problems. Their challenge would be to present an attractive offer while overcoming political resistance to deploying equipment from a Chinese vendor.

Huawei realized that the team led by Bross might not be enough to get past these hurdles. The people running Sprint hadn't worked with him and have few memories now of interacting with him during this period.

Huawei undertook a bold plan to overcome the corporate and political resistance. They started with the launch of Amerilink Telecom, a US-based company that was, in its early press announcements, described as an independent entity, not tied to Huawei or any other equipment vendor and not targeting any one wireless carrier, seeking to "get into all of the Tier-1 providers in the US." Their CEO announced that "if Huawei's one way we can do that, we'll gladly do it." Here was a company that claimed a broad remit—to pursue *all* US carriers with gear from *any* supplier.

Given that assertion, a cynical observer might consider the structure and staffing of Amerilink to be curious, to say the least. The company was ostensibly founded and funded in 2009 with an investment from Bill Owens, former four-star admiral of the United States Sixth Fleet. Former vice chairman of the Joint Chiefs of Staff. Former board member and interim CEO of Nortel.

Owens had been asked to take the helm of Nortel just as the company was collapsing. For years Nortel's operations were such a mess that they were unable to deliver accurate audited earnings—they couldn't produce an annual report on the company's finances—and when investors finally lost patience, CEO Frank Dunn was fired and Owens was tapped to keep the CEO's seat warm for a year until they brought in Mike Zafirovski to replace him.

Owens had taken over during the late days of Nortel's downward spiral and was never seen as causing the spectacular collapse, but he was one in a string of CEOs who were unable to prevent it. Owens was not a technologist, or really even a businessman. Now he was being engaged to serve as chairman of Amerilink, in a role where he would advocate to US policymakers and corporate prospects on behalf of the company, and bring his own credibility, honor, and relationships to bear on its efforts to win business in the US.

The CEO he brought in to run Amerilink, Kevin Packingham, had stood on a podium at the LTE World Summit in Amsterdam a few months earlier, in May of 2010, as Sprint's senior vice president of Product and Technology, to announce the company's plans to deploy a nationwide 4G/LTE network. Packingham was well regarded in the industry and highly respected by his peers at Sprint, and he had been a key person in making purchase decisions on behalf of the company.

His hiring by a vendor hoping to court business with his former employer was not illegal, and few people would consider it unethical. It's a staple for upstart companies trying to get up to speed quickly: hire an insider who will have no trouble walking back into his old suite and making a pitch to the decision makers. Packingham assembled a small cadre of executives and engineers from Sprint's staff and even tapped people from Huawei's in-country team, those best suited to supporting Amerilink as they prepared for a frenetic couple of years pitching business.

However, it was the board that Chairman Owens assembled for Amerilink that really raised eyebrows. Amerilink secured an impressive and unassailable group of American patriots, business leaders, and military officials—people whom any Fortune 50 company would kill to land—and placed them on the board of this pre-revenue technology start-up. The newcomers:

- James Wolfensohn, former World Bank President, American envoy to the Middle East, and advisor to China's sovereign wealth fund
- Richard Gephardt, the former Democratic presidential candidate who had served for years in the US Congress as the House majority leader
- Gordon England, a former secretary of the navy, deputy secretary of defense and the first deputy secretary for Homeland Security

What was the purpose of staffing its board with such an all-star cast? Amerilink's mission was somewhat muddled in the media. As described, they claimed to be pursuing all US carriers on behalf of any equipment vendor, with the mission of helping to sell and distribute the gear to the carriers. But the location of Amerilink was hardly ideal for a company trying to crack open the entire US market. AT&T was based out of Atlanta and Dallas. Verizon operated from Basking Ridge, New Jersey. T-Mobile was headquartered in Bellevue, Washington.

Amerilink decided to base their company in Overland Park, Kansas.

Overland Park is a city of 173,000. It's a town known primarily as the headquarters of Sprint, with employees referring to the massive HQ building on 6200 Sprint Parkway as the Queen Mary on the Prairie. Amerilink secured office space on the fifth floor of a premium office building at 7500 College Boulevard. They weren't *in* the Sprint building, but they were close enough to run into Sprint execs lunching at the new Applebee's across the parking lot.

ONE TARGET

Oddly, Amerilink explained that they also wanted to serve as a company that would vet equipment from foreign countries to make sure it was safe for deployment in US networks. The list of other vendors with such challenges—great technology but sold by a rival state that couldn't be trusted without independent validation—was short. Ericsson was foreign but already trusted and present throughout American networks. Likewise with Nokia, Alcatel, and Samsung. ZTE was the only other major company that met that description, and their own rivalry with Huawei was too fierce for them to contemplate a joint operation.

Each aspect of the new company's existence seemed less likely than the next. And while they informally acknowledged in future communications that they had been created to vet and distribute Huawei gear in the United States, targeting Sprint, they officially claimed no direct involvement with Huawei. Their lobbying firm, the Glover Park Group, LLC, hired within days of their launch, went further, attesting in a signed Lobby Registration Form on August 13, 2010, that no foreign entity "directly or indirectly . . . directs, finances or subsidizes activities of the client (Amerilink)." This assertion, signed by Joel Johnson, a former senior advisor to President Bill Clinton, that Amerilink was not created to serve as an agent of Huawei, seems difficult to reconcile with what was known about their activities. Just prior to funding the company, Owens had worked as a paid consultant to Huawei. And this was not the admiral's first encounter with Huawei, whom he had approached about a possible merger or acquisition to rescue Nortel, back when he was CEO of the company.

The proposed role of vetting telecommunications gear was also curious. The resources and skill levels needed to do this exceed all but the largest, most sophisticated technical R&D organizations. Gordon England, on announcing his new role on the board, described himself as "impressed" by the company's plans to verify the security of foreign-made telecom equipment, although he couldn't explain how the company would be capable of carrying out this vetting, telling the *Financial Times*, "My personal view is that it isn't so complicated."

He was wrong. It was extremely complicated. If the only board member with technology leadership experience was this clueless, what chances did the others have of providing oversight? Industry experts consider the mission of testing network infrastructure for hidden vulnerabilities to be a massive, complex task, debating whether it is merely complicated or virtually impossible. Perhaps Bell Labs, in its heyday with tens of thousands of technical experts across all sectors, had the on-board talent to independently examine a switch with a million lines of code, or disassemble a radio head to identify suspect software or hardware. But no self-funded start-up could claim anything approaching that level of expertise.

Most importantly, Amerilink was hardly independent. Despite any questions of their financial relationship with Huawei, they were explicitly Huawei's leading partner in the effort to win Sprint business, and the internal conflicts of interests were prohibitive. How likely would they be to call out their sole partner in front of a customer and declare that they suspected a device was corrupt, intended to steal information or grant hidden access to communications? These determinations are hardly absolute; when Vodafone discovered worrisome software on their Italian set-top boxes from Huawei, the companies vigorously debated whether this was deliberate malware or sloppy coding, the innocent errors that all companies make in good faith. What position would Amerilink have been in to make that "independent" call?

Undeterred, Amerilink launched an all-out effort to land the Sprint contract.

18

Leapfrog

The threat of having Chinese telecoms systems
inside of American networks . . . presents an
enormous risk, a national security risk . . .
—Mike Pompeo, former US Secretary of State

There was something that the senior people staffing the new teams at Huawei US, Futurewei, and Amerilink had in common: They were almost all close, personally and professionally, to the troubleshooter that the Sprint board had hired to turn the company around, former AT&T exec Dan Hesse. Hesse's strategy to save Sprint involved significant changes to the network, including launching new services and passing the competition to deliver high-speed data. This meant he would have to build out a massive next-generation wireless network based on the newest 4G/LTE technology standard, a project that would deliver a $7 billion contract to the vendor selected.

Joining Hesse at Sprint was his right-hand man Steve Elfman, who served as president of Network Operations and Wholesale, with responsibilities including network build-out, operations, handset oversight, and technology innovation. Hesse and Elfman added some trusted hands from the team that had worked with them for years at AT&T and Terabeam, the start-up they briefly led after leaving AT&T. This was a team that had been around the block, and they were familiar with all the vendors in Europe and the United States, as well as Samsung and LG, two Korean manufacturers who had provided technology to their previous companies over the years. They were not as familiar with Huawei, so if Huawei wanted a piece of Sprint's next-generation network build-out, the Chinese upstart had a long way to go.

Hesse was not unsympathetic to the challenge facing Huawei. When he ran AT&T's equipment business in EMEA, he sympathized with colleague Jim Brewington's efforts to get him to sell AT&T's wireless gear to the European carriers. Hesse explains that the equipment wasn't ideal for the European carriers—they weren't interested in 1G analog gear, they wanted 2G/GSM equipment—and in addition, "they certainly were not going to buy American" if they had good alternatives coming from Sweden, Finland, France, and Germany. Back when the shoe was on the other foot, Hesse tried to get around this reluctance by convincing AT&T to buy up smaller European equipment vendors like TRT in France and PKI in Germany, both of which were second-tier players in the European market. Even those acquisitions weren't enough to transform AT&T's sales into Europe.

Huawei attempted a similar approach to gain US acceptance, pursuing acquisitions of American equipment makers. They had been in talks with Nortel, and they came close to negotiating a deal with Motorola, but both transactions fell through. Other acquisitions were blocked by the federal government, sometimes very late in the game. It was clear that Sprint would be considering Huawei as a stand-alone bidder, going up against Ericsson, Nokia, Alcatel-Lucent, Samsung, and Chinese rival ZTE.

BETTER, BUT . . .

Elfman went to work evaluating the bidders on performance, price, and reliability. Long before the media knew of the full magnitude of the project, he set up a massive competition between the contenders, including field trials of their radios all over the United States. He recalls giving Huawei thirteen to fifteen cell sites to create in Florida, while the competing bidders deployed 4G base stations in other local markets. Meanwhile, Elfman traveled that spring and summer across China, Korea, and Europe, touring labs, meeting management teams, listening to product road-map presentations. As he visited, reports came in from his field teams. Ericsson was way behind with the technology—they didn't even have the radios to operate in the required spectrum. Alcatel-Lucent was also behind, but "on their way" in building a state-of-the-art solution. Leading the pack on technology were ZTE, Samsung, and . . . Huawei.

"Huawei was way ahead," says Elfman, "the right power levels, the right frequencies." More than that, he saw that the Chinese players weren't just compliant with the specs—they were innovating. "The idea for our 4G network, we called it Leapfrog, really came from a visit to Huawei."

Yet that visit paradoxically showed how Huawei had cultural disconnects serious enough to potentially kill a deal. They rolled out what appeared to be the senior team of Huawei, but Elfman couldn't make heads or tails of it.

"They brought the chairman, the co-chairman, the vice chairman, the last CEO, the rotating CEO," says Elfman. "It was very difficult to understand the org chart." Sprint's advance team couldn't brief their execs on the org structure at Huawei, which maintained an opacity that would make the NSA proud but leave a wary prospect confused and ill at ease. It would be two more years before Huawei published even brief biographies of their leadership team.

For a senior executive deciding whether to entrust a "bet-the-farm" contract to a new vendor, the personal relationships are very important. The technical specifications—speeds and feeds—matter, but only if leadership trusts that the vendor is in good hands. This was challenging when Elfman was unable to even figure out who had their hands on the controls.

He did know Admiral Owens, who was present at the meetings in China on behalf of Amerilink, and he met the one person that he knew was ultimately in charge, Founder-Chairman Ren.

"We were treated like kings," Elfman says. "When I met with Ren they had two large chairs that we were seated in, facing the minions. He did most of the talking, working with an interpreter, radiating energy and charisma."

Except for Ren, the team from Huawei barely said a word—they just sat there. At one point, Owens, who was at the table, made a pitch for the Sprint business using a military metaphor, referencing navy artillery. Ren suddenly interrupted, shouting to his American guests, "We will point our guns at Verizon and beat them!" The Huawei executives suddenly came to life, roaring with support as their American guests shifted in uncomfortable silence, trying to process the scene of a former PLA officer exhorting them to join him in crushing America's biggest telecom carrier.

Elfman was careful not to give Huawei false assurances that they were going to win the business. He let them know that he was meeting with their competition and had major concerns about awarding the contract to a company that brought such political complications. They seemed to ignore this, plunging ahead with plans for deployments and doing their best to ply their guests with inducements.

"At the end, they presented me with a gift," recalls Elfman. "They knew that I couldn't accept anything over one hundred and fifty dollars in value, but this jade desk set must have cost thousands. It had my name engraved on it." The desk set was shipped to his home in the US and then forwarded to the company's charitable organization, the Sprint Foundation.

The European vendors tried to offer gifts as well—Hermes scarves, and so on—but they were pikers compared to the Chinese. It all ended up in the foundation. Later, the teams from Huawei and ZTE offered Elfman wine. "They knew I collected wine, so they would deliver a '1982 Petrus,'" he says. Knowing China's reputation for counterfeits, Elfman wondered what was really inside, but he left the corks in and turned the bottles over to the foundation too.

The funny thing is, Elfman really liked the Huawei teams. "We liked talking to them," he says. "It wasn't like a real pressure. They were decent people."

COMMERCE ON THE PHONE

Not everyone agreed. As Elfman was putting the bidders through their paces, the US government was starting to lean on Hesse to avoid the Chinese technology. Newspaper accounts say that Senate staffers met with Sprint representatives in August of 2010 to warn the company about Huawei and ZTE, but not long after that Hesse also got a call from his government team say-ing that Gary Locke, the US secretary of commerce, wanted to speak with him. Hesse took the call alone in his office.

Locke explained the government position. "I understand there are Chi-nese companies that are going to be bidding," he said. "You can do whatever you want, Dan. It's completely your choice. The last thing I would ever do is tell Sprint what you can and cannot do." But Locke also reminded Hesse that the federal government was a customer of Sprint, and emphasized the importance of security considerations in choosing which networks the gov-ernment could rely on.

This conversation was ironic not just because Locke was the US's first Chinese American secretary of commerce but because he and Hesse had a history together. They had toured China years earlier on a goodwill mission when the men wore different hats.

In 2002, Hesse was trying to make a success of an advanced—but bizarre—form of wireless communications. He had left AT&T and taken a job leading a Redmond, Washington–based start-up called Terabeam, which promised high data speeds over a wireless connection using high-frequency millimeter waves and other, even higher frequencies; the company was pro-moting a technology that used light flashes to transmit signals over long distances, from a laser to a sensor, like fiber optics without the fiber. Some termed it "freespace optics."

During Hesse's short tenure there—the technology was perhaps ahead of its time—he had accompanied Locke, then governor of Washington State, on a business development tour of China. All but unnoticed by the media, the two had traveled together on a mission to win Chinese business and strengthen trade ties. Hesse was trying to find Chinese carriers who might be willing to take a chance on this novel wireless technology, and Locke was seeking to build political relationships that would help American compa-nies like Seattle-based Boeing and Microsoft win business in their fastest-growing market. Both efforts were successful, with Terabeam landing a trial at Great Wall Broadband, and Boeing securing hundreds of millions in air-plane contracts.

"Locke was the big China cheerleader when he and I had last been together," says Hesse.

Now, a few years later, Hesse understood that Locke was urging him to *avoid* trading with China. That the US government would strongly prefer that Sprint not put Chinese gear into the network.

But Hesse says he wanted to separate the technical decision from the political: "I wanted Steve Elfman to evaluate the vendors on their merits, not in response to political pressure. I kept him out of the discussions with Commerce."

In the end, Hesse's decisions satisfied the commerce secretary. "I would have made exactly the same decision had I never got a call from Locke or anybody else," Hesse says, citing Huawei's lack of US reference accounts, experienced installation and service crews, and relationships with the American tower companies. But he acknowledges that their technology was good and their prices great. Sprint executives reported much lower prices from China than the quotes coming in from the European and Korean bidders. Sometimes absurdly so.

NO BOTTOM

When the award was decided, the Chinese companies learned they were out, both Huawei and ZTE. But they wouldn't give up. On a multibillion-dollar bid like this, vendors were known to shave prices late in the game to snatch victory from their competitors. They might throw in additional services for free or even cut the "best and final" bid by another ten or twenty million dollars.

Elfman recalls getting a last-minute call from ZTE while on vacation in Tuscany. "I was in my hotel room when I got the call," he says. "They told me, 'We'll give you all the 3G gear for free. For the 4G, we'll cut our price another eight hundred million dollars.'"

Elfman was stunned. They were already the low bidder—so low there was no way they were going to make money on the contract. Now they were offering to cut nearly a billion dollars from their bid. *A billion dollars!*

"I phoned my head of engineering and told him about the call," Elfman says. "I said, 'Next call I get, I'm going to ask them to give *us* seven billion dollars and we'll agree to take their gear.' They just didn't realize that this wasn't about price." (ZTE might have accepted that offer too.)

Both Hesse and Elfman acknowledge, however, that the aggressive bidding by both Huawei and ZTE had an effect on the pricing they ultimately required from the companies that were chosen to supply the gear, Alcatel-Lucent, Ericsson, and Samsung, further reducing their margins and limiting their research budgets. By the time the award was delivered to the winners—the "winners"—their own prices had been pushed so low by the Chinese bids that it became nearly impossible to turn a profit.

Elfman never did reach a comfort level with the security questions. He asked his team what it would take to be able to look at Huawei equipment and say, "This is a security problem!" They let him know that vetting equipment in a demo lab was not a useful way to ensure it wasn't corrupt, especially when Huawei would be giving Sprint the equipment to test.

"These guys were not idiots—there wasn't going to be something in the test gear that would flag a problem," he points out. "The only way to have confidence in the gear supplied was to have confidence in the gear supplier.

As for Amerilink, a few weeks after they failed to land that one big customer, they closed up shop.

19

Sniffing Around the Nuclear Missiles

In the War of Liberation, we continued the policy of
first encircling the cities from the countryside and then
capturing the cities, and thus won nation-wide victory.
—Mao Zedong

Though Huawei had failed to land Sprint, their US sales force did not remain idle. Large deployments—billions of dollars, in some cases—were being carried out by Verizon, AT&T, Sprint, and T-Mobile, and they all seemed off-limits to Huawei. But the United States is a huge country filled with vast, lightly populated areas. For the biggest carriers, it's not economical to deploy cell towers there, so they often leave it to small, independent service providers, operating on shoestring budgets and working with little overhead. Even the equipment vendors tend to overlook these places; Ericsson would rather have one of its salespeople submit a proposal for a hundred-tower project in San Francisco than a three-tower project in Popejoy, Iowa, and the field techs don't have much interest in making a four-hour drive to visit a site that needs troubleshooting.

For years, Huawei has been a boon for these rural carriers, offering cellular equipment at half the price of the US or European rivals, and throwing in free vendor financing and field support. For many of these local companies, the extremely aggressive pricing and comprehensive support make Huawei the only option when they need to deploy a new cell site.

Helped by federal subsidies for rural service, these operators play an important role in providing wireless voice and data service for Americans living remotely, in small towns or on farms and ranches. A former chief of the Wireless Bureau at the Federal Communication Commission offers an interesting take on why Huawei has been so willing to sell into these accounts when there is clearly no economic appeal.

"Huawei understands something about American politics," he says. "Oklahoma has as many senators as California. If anyone ever tried to remove the Huawei gear it would leave those American citizens literally without a lifeline." His words show insight on how the government works, and senators from affected states have, in fact, been pressing the FCC to find alternatives to pulling out Huawei, which would cut off service to rural customers in their states.

But there was perhaps another reason why Huawei was placing equipment in rural towers, practically giving away equipment to customers who presented no opportunity for larger sales. This theory was initially developed not by the FCC—though they would eventually get there—but by a team working on the other side of the National Mall, in the J. Edgar Hoover building.

STEALTHY DEPLOYMENTS

John Lenkart is a West Point graduate who hadn't really used what he learned in his engineering classes until he left the army and joined the FBI. "I niched myself into looking at counterintel operations that were more technical in nature," he says. "Not just chasing spies." Starting around 2009 he undertook an entrepreneurial endeavor, building a team at the FBI to assess and investigate the national security concerns with Huawei's activities in the United States and around the world.

"We could have tried to find the smoking gun linking Chairman Xi to companies like Huawei, but that's a fool's errand," Lenkart says. "Instead, we started with an assumption that Huawei served as a capability to meet the goals of China. If that were the case, what would they be doing?"

As he and his team started looking at the Tier 3 telecom service providers like the rural wireless carriers, they made an interesting discovery. "A very significant portion of DoD (Department of Defense) communications are carried on commercial infrastructure," Lenkart points out. "There's very little that is carried end to end by DoD itself. And if you look at where all these nodes of infrastructure are, whether they're ICBMs, or JSOCs (Joint Special Operations Commands, which house the elite groups like Delta Force and SEAL Team Six), they're in nowheresville." In other words, the US government isn't locating intercontinental ballistic missiles in Palm Beach, Florida. Most of the secure bases and classified operations are in rural America.

When Lenkart and his team overlaid a blueprint of these locations on top of the companies and towers that use Huawei gear, they were stunned at what they found: Huawei's deployments in rural America mapped inexplicably close to the areas of sensitive military and government operations.

In November of 2019, FCC commissioner Jessica Rosenworcel delivered unclassified testimony about the situation. She described the desolate area surrounding Malmstrom Air Force Base, where nuclear missiles sit ready for launch. Malmstrom is on the edge of Great Falls, Montana, and one hundred miles from the nearest . . . well, one hundred miles from anything else. Montana is "Big Sky Country," with lots of land, few residents, and the third-lowest population density in the country. It does not present an appealing market opportunity to any of the major mobile carriers.

Overlooking the ultrasecure facility—home to more than a hundred intercontinental ballistic missiles—is a cell tower operated by Triangle Communications, an independent, locally owned operator and the only wireless network based in this rural area. They advertise, "We build towers where the big guys won't." The provider of the equipment in their base stations? Huawei.

The federal government bars its employees from using Huawei handsets and blocks the large carriers from putting Huawei's gear into the network. But for the missile control officers living in or near this ultrasecure facility, wireless communications are passing through Huawei boxes. Can China use the equipment to listen in on the communications at the base? There's no way to know for sure, but as the company controlling those radios and switches in the cell tower, they may be able to shut down service in a time of crisis or launch denial of service attacks on the region. Even if the communications are encrypted, a bad actor could use metadata to glean information about who's on the base that day, who's connected to the tower, who they're communicating with, or how much traffic is flowing, to gather vital insight about base operations.

There's more reason to be concerned about the presence of China's leading technology firm on the edge of one of the country's most secure nuclear weapons facilities. As Commissioner Rosenworcel continued, "This is just one military base in Montana. But there are others like it."

GIVE 'EM A MEDAL

This isn't just idle fearmongering. The people charged with protecting America against foreign and domestic threats have dug deep to understand the extent of the violation. As part of the FBI's counterintel investigation, Lenkart and his team met with many of the rural service providers using Huawei gear and with the competing US equipment vendors. He was surprised at what he learned: "The other equipment vendors told us, 'If we sold our equipment under the terms of this deal, the sun would run out of fuel before we would get our money back.' So, what's the value? The value is to the nation-state of China. It's not a business deal. It's a projection of geopolitical goals."

The FBI redoubled its efforts to brief telecommunications carriers on the dangers to American security of putting the equipment in their networks, but have had little effect on their decisions. "To say, 'Do it for America,' it falls in deaf ears," Lenkart says. "Understandably. They have to think about shareholders and owners." And he delivers this kicker: "Huawei understands that the rural carriers can get a federal subsidy to put in cell towers. The US Treasury has to float bonds to pay for it, borrowing the money from China. The carriers use the money to buy gear from Huawei, who uses the gear to potentially degrade and disrupt communications at our military bases. They got us *three times*," he groans.

"I want to go to Beijing and give the guy who came up with this a medal."

20

Everyone Has a Price

Huawei is a symbol of the national future of China. The
people know, the government knows, that if Huawei cannot
survive, this country will have no hope for rejuvenation.
—Yan Xuetong, dean of the Institute of International
Relations at Tsinghua University in Beijing

The access to remote secure facilities may have given some advantage
to Huawei and China, but they remained frustrated in their efforts to
break open a bigger piece of the developed world's networks. In their normal course of business, Huawei did what all companies do when entering
a foreign market: they hired people who had a familiarity with the local
business environment.

It's hardly a novel idea to recruit people with good reputations and
personal relationships within a targeted market. Global companies know
there's a value in hiring people who have worked with the local regulatory
officials, people who understand how the game is played. There's no substi-
tute for someone who's been in that market and even has held a position on
the other side of the table.

But as with so many things Chinese, the scale was different, especially as Huawei ramped up their efforts to win access to the countries still resisting their entry to the market. Coming from a national culture that places enormous value on connected insiders—*guanxi* refers to networking and relationships, yes, but also has connotations of gift-giving, exchanging favors—they undertook a massive effort to recruit and retain leading political and business figures from all over the world to advocate on their behalf.

AIMING HIGH

China made no secret about hiring the most influential, respected members of the American House of Representatives, Senate, and even the intelligence and defense departments to advance its case. In some instances, the CCP made them its spokesmen. And in the most astonishing moves, the people it recruited and hired were the very ones who had led the charge to resist Chinese infiltration to American telecom markets.

In October of 2010, shortly after Huawei was effectively blocked from deploying its gear in the Sprint 4G rollout but before the outcome was known, four members of the US Congress sent a letter to Julius Genachowski, chairman of the FCC, urging him to investigate Huawei and ZTE. The letter raised several urgent concerns, including the fear that putting their gear in our mobile networks "may create an opportunity for manipulation of switches, routers, or software embedded in American telecommunications network so that communications can be disrupted, intercepted, tampered with, or purposely misrouted." The presence of gear from these companies, they concluded, "would pose a real threat to our national security."

The letter was signed by three senators and one member of the House of Representatives, but the heavyweight was one of those senators, Joseph Lieberman, chairman of the Senate homeland security committee and the author of the legislation that created the Department of Homeland Security. He had also been nominated to run as vice president on the 2000 Democratic ticket.

The damage the letter dealt to China's efforts was great. In a spirited defense of ZTE, a few years later, one of their paid advocates described how he was working to find out what can be done to "raise the level of trust in ZTE."

That spokesman? Joseph Lieberman.

• • •

The letter from the congressmen had asked other questions like, "Does the FCC work with the Department of Homeland Security or the Intelligence Community to better understand the potential risks posed to US telecommunications networks?" If the FCC had been doing such partnering, they might have worked with the lead cyber expert at the DHS, Andy Purdy, the White House national security staffer who drafted the *National Strategy to Secure Cyberspace* and went on to launch and eventually lead the National Cyber Security Division at the DHS. Purdy worked hard to protect America from the threat of Chinese intrusions, and he was the lead cyber official for the US government, specifically responsible for protecting our networks from foreign infiltration and manipulation. Huawei objected to the way people like Purdy characterized them and found that the policies he advocated made it harder for them to do business in the US. Huawei retained experts to counter his arguments, and paid to put them in front of the American public.

One particularly enthusiastic and combative advocate appeared on CNBC's *Squawk Box* news show on February 19, 2020, during the heat of the debate about the threat from Huawei. The spokesman decried what he called the "campaign to carpet-bomb Huawei out of existence," and mockingly referring to "Huawei Derangement Syndrome." This hurt American companies, argued the advocate, Huawei's new chief security officer, who had had been hired by the company about a year after the Lieberman letter. He insisted to the American people that there was "no national security reason" to justify blocking China's largest telecom-equipment maker.

This fierce advocate for allowing Huawei into American networks? Andy Purdy.

• • •

It would be impressive enough for Huawei's recruiting efforts if the list ended there, but these hires were not even the most connected or influential. Many of these individuals weren't just being recruited to be marquee names, like the board members at Amerilink. These people were *on the payroll*, going to work every day to bring Huawei into markets that were wary of the risk. In this game of cat and mouse, it became hard to identify which people were trying to guard America's networks and which were trying to assure the guardians that there was nothing to see here.

Samir Jain, for example, served on the National Security Council as President Obama's senior director for cybersecurity policy, having previously been associate deputy attorney general at the Department of Justice, where he took on China, securing its commitment not to engage in intellectual property theft for commercial gain. He described his key role as helping draft Obama's executive order blocking the property of companies or countries engaged in malicious cyber activities. This 2015 order was considered a gloves-off response to the Chinese threat, declaring a "national emergency" and enabling a "whole of government" response to cyber threats, including allowing the Department of the Treasury to use its property-blocking authority. Legal commentators at the time found it notable how strongly the president's order "flexed its economic muscle" when it came to dealing with China.

Having left the National Security Council in 2017, Jain took his Harvard law degree to Jones Day, the prestigious DC law firm that counts Don McGahn, President Trump's first White House council, as a partner, and has provided more than a dozen senior lawyers to key roles in the Trump White House. One of his first big client wins after joining the firm?

Huawei.

In April of 2019, President Trump directed his Twitter feed against the news that Huawei had retained Jain as a lobbyist. "This is not good, or acceptable!" raged the president, on learning that one of Obama's top cybersecurity officials would be working for the company he had worked so hard to rein in.

"Money talks," said Jain's former DHS colleague Nate Snyder, speaking to *The Hill*, "and from that perspective, he's probably making a pretty good paycheck . . . From a national security perspective . . ." Synder went on, "we've got a big problem here."

SHIFTING ALLEGIANCE

In other countries, Huawei didn't aim any lower.

In June of 2019, Huawei sent a champion to face down the British Parliament that was becoming increasingly alarmed about reports that the company's network gear presented a threat to national security. The equipment maker's lone representative was regaled as a hero to the masses across China watching the hearing online as he stood up to questioning from members of Parliament. At one point, Julian Lewis, a Conservative member, asked the representative about the law in China that requires companies to cooperate actively with the intelligence services. The Huawei representative was defiant and confident as he said, "There are no laws in China that obligate us to work with the Chinese government on anything whatsoever."

If the members of the Parliament seemed deferential, at least at first, to Huawei's advocate, it's understandable. Until he took the role of Huawei's global cybersecurity officer, John Suffolk had served for years as the UK's chief information officer, advising these same members of Parliament on issues of cybersecurity and national protection from foreign threats.

Now, he would need to rely on their deference, as his answer was clearly misleading, at best. Even in the freest societies, there are laws that compel companies to cooperate with the home government. In the United States, communications companies are required to make sure their gear can be used for what is termed "lawful intercept," where a court order allows the government to tap a person's phone. In China, the constraints on government intrusion are negligible, and legal warrants need not be issued to permit the CCP to monitor a citizen. Article 14 of China's National Intelligence Law *explicitly* requires Chinese companies to cooperate in intelligence gathering both in China and around the world.

Suffolk may have been a useful hire for the Chinese company, but they didn't stop there. Huawei pulled a coup when they named Sir Mike Rake to Huawei's UK Board. He was the former chairman of British Telecom, ex-head of the Confederation of British Industry, and advisor to the prime minister. Sir Mike used his new platform to caution that restrictions against Huawei equipment would hurt the UK's efforts to deploy broadband, an area that he recently held great responsibility over.

The list of prominent leaders taking money to represent Huawei's interests is long, including Lord John Browne, former CEO of British Petroleum, and Andrew Cahn, the former UK trade minister.

If China were not content to see a mere *advisor* to the prime minister joining the team, the most brazen move came in support for the Belt and Road Initiative (BRI), the country's massive program to develop ports, highways, and other infrastructure in furtherance of Chinese influence around the world. When it wanted to win more support from the United Kingdom, it went right to the top and hired the former prime minister himself, David Cameron, only months out of office after having served six years leading the British government. On the Chinese payroll, Cameron oversaw a $1 billion development project that represented a mix of public and private companies. His responsibilities in the initiative were vague, at least as described by his spokesman, who said Cameron would "play a role in a new UK–China bilateral investment fund that will invest in innovative and sustainable growth opportunities in both the UK and China to create jobs and further boost trade links . . ."

But given the history of China's BRI in advancing Chinese political interests around the world, the former head of state for the UK faced the possibility of having to represent a foreign government's interests ahead of his own country's.

KNOW, OR SHOULD KNOW

Is this behavior unethical? Is it unpatriotic? And does it present real risks, given the knowledge that these people possess before signing on to advocate on behalf of the "other side," against which they had fought for years?

"The revolving door is a hard problem to solve," says Ed Freeman, a philosopher and professor of ethics at the Darden School of Business at the University of Virginia. The author of numerous books on business ethics and stakeholder management, Freeman teaches students and executives how to examine the dilemmas they may face in business when confronted with choices that involve issues of personal integrity and obligations to shareholders, colleagues, customers, or society at large. And he sees this behavior as potentially troubling. "You'd like to think people with integrity would take those jobs because they think they can do them with integrity. And sometimes that's a bit naïve because they don't realize the difficult positions they're going to get put in."

So does he think the people recruited to front for Huawei don't realize they might be asked to advocate actions that run counter to the interests of their own country? "Maybe they authentically believe in their hearts that there's nothing to worry about here," he acknowledges. "But these are not amateurs."

The problem lies when they claim plausible deniability; these people can only be expected to work with the information Huawei shares with them. But, as Freeman notes, they can't be confident about what is disclosed to them as a spokesman or advocate: "At senior positions, the standard ought to be if they knew or *should have known*" that something they were fed from their employer was not what it seemed to be.

What about the argument that everyone deserves to be represented, like a defendant in a trial? Lawyers can't be faulted for taking on unpopular clients—it's a requirement of the profession to give the best representation to your client possible.

Freeman doesn't see it that way. "It's a different system," he observes. Lawyers are expected to use the existing law to make the most vigorous case possible for their client.

His point is an important one: No lawyer presents a case that is based on his own personal integrity and past deeds, asking the jury to accept his arguments because of his own contributions to society. Lawyers work on facts and precedent of the law. Advocates like the ones Huawei hires bring personal history and relationships and seek to transfer them to their "client." They are in fact using their own accumulated capital—a reputation for integrity, a track record of protecting their home country—to wrap Huawei in a protective barrier. They are selling their own integrity, not just their time or ability, and, unlike with a lawyer, if it turns out their client really is up to no good, it *is* on them.

21

The Debt Trap

If you owe your bank manager a thousand
pounds, you are at his mercy. If you owe him
a million pounds, he is at your mercy.
—John Maynard Keynes, English economist

If securing influence in Western, developed countries was a simple matter of money, gaining clout in the developing world was no harder, and Huawei flexed its muscle when it came to sweetening the pot for—or putting pressure on—prospective customers. Many of them were already struggling with the broader societal ills that come from insufficient funding for national infrastructure like transportation, water, power. This is generally considered to be beyond the kit for a company selling base-station radios and servers, but Huawei was able to take its game up another notch.

NO RESTRICTIONS, BUT A FEW STRINGS

Bundling multiple products and services together isn't unusual, with many companies tying software and support to the sale of network hardware, but Huawei's bundling is on a level that most Western telecom executives couldn't imagine. Pat Russo, CEO of Alcatel-Lucent, remembers the reports she was getting from her field sales leaders who were getting killed in their bids against Huawei in Africa and Latin America. Alcatel-Lucent would submit a bid to the national carrier of a developing country. Then, she says, "the Huawei bid would come in, not only at a fraction of the price, but wrapped with promises from the Chinese government to provide development loans that could be used for bridges, clean water, highways. They would bundle major construction efforts along with their communications solutions, and they would make the rest of the help from China contingent on them buying the communications equipment."

This government assistance changed the game entirely; the vendors may have been battling it out on product quality and price, but the customers, often national telecom ministries, had broader desires that these infrastructure offers met.

To the extent that public aid from the home government of a vendor is helpful in securing a contract, the United States should have had the ability to equalize competing offers of assistance from the Chinese government. This did not turn out to be the case. While the US government never ties a foreign aid project to the selection of a US vendor, the expectation of future assistance should at least have a subtle influence on customers in developing countries. But as American and European companies learned, Huawei had another advantage: projects supported by the Chinese government came with few constraints or demands.

A banker in East Africa, who wishes to remain anonymous, explains why these nations often favored Huawei: "The US would say, 'We'll pay to build a pipe so people in the village don't have to walk to the river to get water.' But it would include rules: 'You can't use child labor; you need to give breaks every so often; the water must be treated to make it safe.' The Chinese come with no rules. It brings the money, the pipe gets built. Maybe the water isn't so safe, and some people will get sick, but people get water in their homes fast. The government doesn't have to wait."

NOT A BANK, AGAIN

Huawei had another trick in their bag. Vendor financing, so important to winning deals in the go-go nineties, had all but dried up after the telecom bust, but for Huawei the party was just getting started. If deals like the Winstar loan seemed absurd, Huawei took the practice to a new level, one that couldn't be justified in financial terms.

To illustrate, the loans made by US-based vendors were granted on an expectation that the customer would be able to pay them back. Whether it was a subsidized loan through the federal government's Export-Import Bank (EXIM) or a loan financed by the vendor themselves, the real effect was to reduce the interest rate the customer would have to pay, help ease their budget pressures, and in effect *improve* the odds that they could pay the money back. A private company extending credit has no interest in securing control or political advantage over a deadbeat customer; they just want to get paid back. No maker of cellular radios wants to be repaid with collateral by seizing a wire-pulling factory in Somalia. Even the EXIM bank is interested only in helping a local company close the transaction.

China's state-owned banks, by providing massive loan agreements to Huawei, were able to virtually guarantee that carriers with limited budgets would look exclusively to that vendor as the supplier of choice. This was a winning combination for Huawei: Low prices and aggressive government support, both through lending and tied foreign aid, were enough to start winning significant market share across the developing world and even in Europe.

Critically, the Chinese state insulated Huawei from the catastrophic losses that the company could face if the vendors failed to repay, as they often did. As we saw, after Lucent extended over $7 billion in vendor financing during the nineties, often to undercapitalized carriers who took the equipment and never paid back the loans, the impact on Lucent's finances were devastating.

For Huawei and China, the consequences of failing to pay back a loan would have more sinister implications for the borrowing phone company, especially across the developing world, leaving a trail of national telecom carriers beholden to their vendor, an odd about-face from the traditional power relationship.

China began to use the defaults by Huawei customers as a "debt trap," a deliberate method to turn prospective customers for telecom gear into compromised borrowers, according to many analysts. By defaulting on their loans to China, these companies all but turned over ownership of their companies—and countries—to the CCP.

The impact on those customers has been devastating. A telecom executive in Ethiopia, who does not want his name used, describes the world he and his colleagues operate in: "Our network is almost entirely from Huawei—them and ZTE. We don't know if our calls on our own networks are being listened to, but even if we find they are, we don't have the muscle to fight back." As he explains, they owe so much money for the equipment that Huawei effectively owns their network.

He continues with a story from a few years ago: "Internal discussions were underway about privatizing Ethio Telecom," a state-owned telecom service provider in Africa's second-most-populous country. "Huawei just walked in and presented a detailed valuation of the company, including details on usage, customer base, internet traffic. This was data that even we didn't have. They were apparently planning on making a bid for the company," he says, "and they were shameless about showing what information they had pulled from *our* network operations center."

In this case, buying out the carrier may not be necessary. With a total of $3.1 billion owed by Ethio Telecom to the Chinese government and little chance to repay, the lender may end up calling the shots anyway.

There is an irony here: The money was spent to upgrade Ethio Telecom's operations. And the result? According to a recent article in the *Africa Report*, Ethio Telecom's service is ranked 170 out of 176 countries.

The danger of this debt trap is well understood even if a solution is elusive. Privatization may be a path out of the control wielded by China, but even that path is blocked by the debt: Free cash flow (the best measure of the value of a company like Ethio) is dramatically reduced by the need to service the debt to Huawei, estimated at nearly a billion dollars in interest payments over the life of the loan. The consequences are grave, going well beyond the ownership of a company. As the *Africa Report* puts it, "Given Ethiopia's indebtedness to China, there may well be a risk to long-term national sovereignty."

This debt trap would turn out to be only one of the country's—and the continent's—problems when it came to Huawei.

Hacking a Continent

No one has ever found anything that the
adversary has successfully hidden.
—Old intel saying

In *The Art of War*, Sun Tzu shared his guidance on the value of espionage in displacing the massive costs of raising and deploying an army. He described how a nation can improve the chance of success over an adversary by securing "foreknowledge," and assailed those who "remain in ignorance of the enemy's condition simply because one grudges the outlay of a hundred ounces of silver . . ."

It's hard to estimate the value today of a hundred ounces of silver in 500 BC, but the $200 million China spent in 2012 to build and equip the African Union (AU)'s headquarters is certainly more than that. Still, it was money well spent, as China worked to extend its influence in the developing world and sought foreknowledge of what was being discussed, planned, and negotiated by the political and business leaders of the continent. It would be five years before regional experts and the European press learned how it may have been securing that foreknowledge, according to investigations from France's *Le Monde*, in a story that has been little reported in the US media.

BEWARE OF SHENZHENERS BEARING GIFTS

In 2012, when the Ethiopian prime minister cut the ribbon on the gleaming new office and meeting complex, he thanked China for its selfless gift to the people of Africa and proclaimed that "the future prospects of our partnership are even brighter." He praised the Chinese government for its "commitments for a win-win partnership . . ."

He was half right.

China had been eager to engage in the scramble for Africa and win a stronger position in both selling to the continent and securing access to resources like rare minerals and energy. Its approach, according to multiple investigations, was less charitable than it appeared, and it included an intent to secure compromising information on the continent's decision makers.

From the start, China controlled nearly every element of the center, located in Addis Ababa, the capital and largest city of Ethiopia. That was a benefit that came with its offer to fully fund its design, construction, and outfitting, right down to choosing the swivel chairs that would go into each office. With a commitment of $200 million, China bought a lot of freedom and flexibility as to how the complex was set up, including installing and configuring the data center, which contained the secure servers that would handle the massive amounts of commercial, political, and military information that is managed by the African Union. The AU center was designed, equipped, and managed by Huawei, with Huawei-trained local technicians on site to oversee operations. Huawei was selected as the vendor for the communications equipment in the data center, both for internal and external use, while ZTE picked up some of the building's other communications needs. Staffed with engineers and architects imported from China, the project moved quickly as the twenty-story, half-million-square-foot complex rose over "the political capital of Africa."

In January of 2017, five years after Ethiopian prime minister Meles Zenawi inaugurated the building, a technician working in the data center made a surprising discovery. Reviewing traffic logs, he found that every night between midnight and 2 AM, traffic on the data center's servers was spiking.

Sources told *Le Monde* they had discovered that the system had been installed with back doors that granted access to the system and its controls. They claimed that sensitive information on business and trade negotiations, military planning, and political considerations were being forwarded nightly to servers in Shanghai. According to investigators, the spying had commenced with the opening of the center and continued unseen for five years.

The embarrassed administrators of the AU headquarters told *Le Monde*, "This has gone on too long . . . We have taken steps to strengthen our cybersecurity, a concept that is not yet a habit among our bureaucrats or heads of state. We remain exposed." According to press reports, they thanked the Chinese engineers for their work and sent them packing, avoiding any public scandal or reckoning with the Chinese benefactors as they began the process of remediation. After forklifting the racks of Huawei servers out of the facility, the center was completely re-outfitted with gear from other vendors.

In Mandarin, *fang si* may be the closest equivalent to the Yiddish word *chutzpah*. Whatever the best translation is, the Chinese demonstrated it when, after the African Union re-outfitted the center with another vendor's servers, Huawei offered to provide engineers to configure the new data center.

Not surprisingly, this offer was declined.

Since the discovery, the wireless systems that were deployed to facilitate video conferencing have been replaced with cabled connections. All communications are now encrypted and no longer pass over the public Ethiopian network. And yet, despite the extensive efforts to purge the surveillance technology from the center, there was one more discovery made. In preparing for the July 2017 AU summit, just six months after the discovery that the servers were delivering information to Shanghai, a team of Algerian experts was brought in to scour the building and search for any more breaches. According to investigations by *Le Monde* and the *Financial Times*, their room inspection uncovered numerous microphones placed in walls and hidden in offices. It appeared that every conversation between heads of state, military leaders, and business heads was susceptible to eavesdropping by those who had constructed and outfitted the building.

The whole affair left the officials of the African Union frustrated at their excessive trust in their Chinese benefactors and their inability to secure their own communications from prying eyes. As one official commented, "Everyone seems fine that we're a sieve. We let ourselves be listened in on, and we didn't say anything."

An investigation by Danielle Cave of the Australian Strategic Policy Institute confirmed Huawei's central role in designing and deploying the secure communications center. She found contract announcements, listed at the time on the Huawei website and now mostly deleted, that trumpeted the cooperation between China and the African Union: "[Huawei's] solution deployed all computing and storage resources in the AU's central data center where it seamlessly connects to the original IT system."

Whoever came up with the scheme must have missed Sun Tzu's twice-repeated admonition on how to use spies to secure knowledge, when he commanded the reader to "Be subtle! Be subtle!" Perhaps they didn't need to. There seemed to have been little, if any, consequences.

China's ambassador rejected the claims that any such incident had occurred, calling them "ridiculous and preposterous." And it's possible that the story is false; the African Union never went on record acknowledging the hack, and the media investigations relied on numerous anonymous sources at the African Union. But the AU went to considerable expense and effort to replace the data center using non-Huawei equipment and refused the offer of free Chinese assistance in configuring it.

More tellingly, the Chinese did not point the finger at others, like MI6, the Russians, or even the CIA as the real culprits. One would assume that if the Huawei gear were compromised and China and Huawei were not the responsible parties, China would have aggressively pursued an investigation of who besmirched its reputation and brand. It did not. Instead, the issue was brushed aside by the African bureaucracies, and the very claim of a breach denied by their Chinese benefactors.

A Huawei spokesman told the BBC a year later, "If there was a data leak from computers at the AU's headquarters in Addis Ababa that went on for an extended period of time, these data leaks did not originate in technology

Left: Axel Boström, Ericsson CEO, left, in his Stanley Steamer, "mobile phone" in trunk, c. 1903. (With permission, from Artur Attman et al., *LM Ericsson 100 Years* [L.M. Ericsson, Stockholm: 1997] and from the archives of the Centre for Business History) *Right:* Some of AT&T's (and later Lucent's) Bell Labs' many Nobel Prize winners. Drs. Shockley, Bardeen, and Brattain, inventors of the transistor; Drs. Penzias and Wilson, discoverers of the Big Bang and the origins of the Universe. (Reused with permission of Nokia Corporation and AT&T Archives)

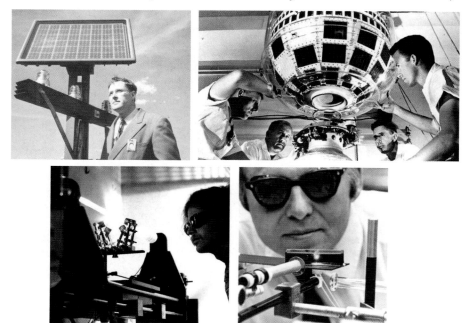

Some of the inventions Bell Labs helped bring to the world: the solar cell, the communications satellite, fiber optics, the laser
(Reused with permission of Nokia Corporation and AT&T Archives)

AT&T leadership, President Bob Allen, center,
inspect first factory in Hong Kong.

Left: Jim Brewington, former head of AT&T Wireless Services.
(Courtesy Jim Brewington) *Right:* Marty Cooper making the
first public cellphone call in New York in 1973.
(Courtesy Martin Cooper)

Alcatel-Lucent Chairman Serge Tchuruk with
CEO Pat Russo, Paris, 2008. (REUTERS/Alamy Stock Photo)

Huawei's Ren Zhengfei, right, takes China's Chairman Xi Jinping, left,
on a tour of a Huawei facility, 2015. (REUTERS/Alamy Stock Photo)

Huawei's research lab, dubbed the "White House," in Shenzhen. (Hector Retamal/AFP via Getty Images)

Huawei's new Dongguan HQ, some of the twelve sections built to represent twelve great cities in Europe. (Clockwise from top left: Peter Stein/Shutterstock, Peter Stein/Shutterstock, fansquaresss/Shutterstock, Camera_Bravo/Shutterstock)

A 5G police robot drives on a sidewalk in China. The robot warns people if they are not wearing masks, checks their body temperature, and scans their face to determine their identity. (atiger/Shutterstock)

Left: Video surveillance system in front of the Heavenly Gate to the Forbidden City, Tiananmen Square. (vvoe/Shutterstock) *Top right:* Uighur reeducation by Xinjiang Bureau of Justice WeChat Account. *Bottom right:* Blindfolded, shackled Uighur Muslim Chinese, being sent to "reeducation camps." (Screen grab from video posted anonymously on YouTube. As seen on BBC and the *Guardian*.)

Federal Communications Commission Radio Service Licensure of Huawei and ZTE Customers

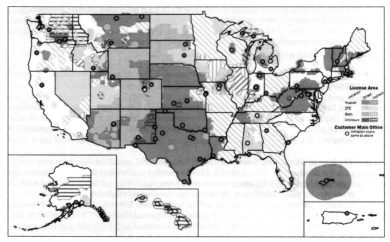

As of September 15, 2020, analysis identified seventy organizations with current or historic relationships with Huawei or ZTE. The combined license area of these organizations covers approximately 86 percent of the United States by land area.

Chairman of Huawei standing in front of an image of a Chinese fighter plane as he addresses Huawei management in 2020, responding to US blocking chip sales to the company, calling it a "pernicious attack" and urging combative response. (Noel Celis/AFP via Getty Images)

A commercial 5G phased-array base station antenna. (Courtesy Taoglas)

China's new phased-array radar system, capable of
tracking and targeting high-speed missiles.

Mercedes 5G automated factory.

5G farm drone.

supplied by Huawei to the AU. What Huawei supplied for the AU project included data center facilities," he said, then oddly added, "but those facilities did not have any storage or data transfer functions." It's hard to imagine the value of data center facilities that can neither store data nor transfer it. What function *did* they perform?

ON MUTE

Why would the phone companies across Africa refuse to acknowledge, on the record, the abuse they are believed to have suffered at the hands of the Chinese government? One reason may be the huge loans that were provided to Huawei's customers, according to the *Wall Street Journal*'s investigation of Huawei's government aid. Those loans were often made to companies in amounts beyond what a commercial lender would have considered prudent. There was little other benefit that came from Huawei's financing offer; otherwise, the carrier could take out a commercial loan through normal channels at similar rates and terms.

Remember the debt trap?

When you consider the costs of empire building, with the billions the Soviet Union poured into its attempt to conquer Afghanistan, or the costs of softer global influence, with the tens of billions of dollars the United States spends on foreign aid and development programs, it seems downright cheap for China to loan money to a country that can't pay it back. Some of the money loaned comes back immediately to the national champion, Huawei, in the form of equipment purchases. And then, as the customer struggles to make payments, the state has the opportunity to use this difficult situation to assert its national interests over the deadbeat debtor.

The telecom debt trap provides a modern extension of power as China secures means to control, observe, or block not just communications but all national activity in its debtor customers. For those who argue that Huawei would not risk being caught compromising their own customers—"If it ever got out, they would be ruined!"—the African Union hack is a ringing counterpoint. According to multiple sources, it happened. But the customer, representing a continent of over one billion people, was unable to stand up to the might of the Chinese vendor/lender/patron, and has remained officially silent, with only leakers and whistleblowers attempting to tell the world.

Today, if you looked out the windows of the AU building, you would see a city under construction, with Chinese companies (often using imported Chinese workers) completing the highways, skyscrapers, and metro system. One job site, about four miles south, might stand out. On December 14, 2020, Ethiopia's minister of Labour and Social Affairs hosted the groundbreaking ceremony for the new Africa Centres for Disease Control and Prevention headquarters building.

The complex is being built thanks to the generosity of China, which is funding the $80 million project. The bidding process is complete for the first phase of construction; the winner was China Civil Engineering Construction Corporation. There is no word yet on who will be providing the data center for the new facility.

The day after Africa's CDC broke ground, Reuters broke the news that AU's technology staffers had discovered that "a group of suspected Chinese hackers had rigged a cluster of servers in the basement" of the AU headquarters complex and were monitoring surveillance video of conference rooms, offices, and other locations at the building. The breach was carried out by a hacking group nicknamed "Bronze President," according to the internal AU memo cited by Reuters.

23

Bugs in the Walls

There are just two kind of companies: those that
have been compromised, and those that still
haven't realized they've been compromised.
—Dmitri Alperovitch, American computer security executive

The described attacks on the African Union headquarters may seem clumsy and obvious. If it seems unlikely that a company with a sophisticated cybersecurity capability would fall victim to such a hack, Nortel's experience shows otherwise.

The business relationship between developed Western countries and China is seen as less mercantile than Africa's relationship with China, closer to parity in the economic power of the parties. Western companies seek the low costs and broad technology portfolios that China offers, and China wants access to the large markets of the more developed countries. There is no question that state-sponsored spying takes place, but developed countries have better cybersecurity protocols, more sophisticated intelligence agencies, and are better able to discover and prevent misbehavior. Does that mean China dare not attempt a hack, or does it just mean it must be more meticulous and work harder to create deniability?

Intelligence gathering is a part of business; you'll find it to some degree in every company in every country. Corporate espionage happens in the United States, but generally at a much lower level of intrigue than in some parts of the world. Nearly every company studies press releases, hiring announcements, and published patents of their competitors. At more aggressive, typically smaller companies, a sales team may send a staffer or hire a private investigator to hang out in a bar frequented by competitors and try to overhear them discussing pricing or strategy for a major bid. A scrappy startup may even "dumpster dive" to retrieve from the trash a competitor's plans that have been discarded by a careless engineer. In a gray area, ethically, but legal in most places.

Big public companies rarely engage in anything like these more aggressive tactics, at least not as a matter of corporate policy, knowing that the consequences of being found out would be harmful to their brand and public standing. People can work a whole career in the United States without ever using or falling victim to these schemes. More typically, large public companies in America seeking useful information may hire talent away from competitors and push the bounds of confidentiality agreements, letting the courts decide if this more genteel approach presents a civil, not criminal, violation.

This code of conduct is not the same around the world. In Europe, companies have been known to provide conference rooms or communications links to rivals that are outfitted with listening gear. They use their national intelligence agencies to procure commercial information from overseas competitors to provide business advantage, something that is explicitly outside the charter of American spy agencies. Germany and France have been aggressive in this regard, as have Asian countries like Korea and Japan, where it's not unheard of for a company to rely on state security services to gather intel on a competitor or even a partner.

But as with so many things, the issue with China is one of scale. And it tends to operate so boldly that it seems to be playing a different game from the others. So it seems with the events that unfolded at Nortel.

GO BOLDLY

At one time, Nortel was the largest, most successful technology company in Canada. They led the world in the development of digital telephone switches and created cutting-edge wireless technologies that enabled broadband data to be delivered over huge distances. Nortel was an early pioneer in technologies that became central for 5G capabilities. Their labs were considered among the finest in the world, second only perhaps to Bell Laboratories.

All of this made them a target for companies seeking to advance their own intellectual property.

On a spring morning in 2004, Brian Shields had settled into his chair at Nortel's massive US campus in Raleigh-Durham, North Carolina. The complex was a rabbit warren, a converted factory site hosting eight thousand employees spread over dozens of buildings. This satellite site included some of Nortel's global corporate security staff, who made sure the company's computers weren't being hacked and the intellectual property was safe. That day, the company would need Shields for both.

Over more than a decade, he had built a reputation for himself as a stickler for improved security. He supported the company's efforts to enforce two-factor authentication, which requires an employee to both use a password and carry a small electronic "secure ID key" in order to remotely log in to the company's network. Such a policy might have averted the disaster that was about to unfold. Shields's persistence would lead Nortel's next CEO to describe him as a man known for "crying wolf." But by the end of the day, Shields would determine that not only had a wolf entered the Nortel network, he was sure it was on the CEO's own laptop.

The first phone call came from Shields's boss in Dallas, Randy Calhoun. A senior executive's password had been used to download a large amount of corporate information, but the executive expressed surprise when he was asked if he had any questions on the material—he claimed he hadn't accessed the information at all. As Shields began his investigation, he found no sign that the executive was the one who had logged in. Searching further, Shields found that the IP address of the computer used to access the data was not that of the executive, but of another, more junior, employee. Finally, checking the "source IP," which should have shown a local Ottawa resident requesting the transfer, he discovered that the information had been moved through an internet service provider outside their own system. "Wait a minute . . ." he remembers saying to himself. "That ain't right . . ." It became clear what had happened.

"We've been hacked!"

Shields immediately called his boss, who assigned half a dozen people to join the inquiry. Teams in Boston, Dallas, and Ottawa were pulled into the investigation. Seven laptops were quickly identified as compromised, but the team couldn't find keystroke loggers, which are commonly used to steal passwords. Most keystroke loggers can be detected with antivirus software, and Nortel was using the best. Whoever had infiltrated the system was working at a deeper level. The malware placed on these computers was virtually undetectable and effectively unremovable.

As Shields continued his investigation, he found something that surprised and disturbed him: He could wipe an infected computer—completely reformat the hard disk—and within five minutes of rebooting, that computer would be corrupt again, with some outside party once again having control over it. That meant the hack was a "rootkit" attack, one that was invisible to malware detectors and granted the intruder the power of a system administrator—"root" control over the system. It was so comprehensive it suggested a state actor was behind it. But which state?

COMPROMISED

When he left for home that night, Shields was consumed with getting on top of this attack and cleaning the system. He knew Nortel would also need to develop a solution to maintain security in the future. But he was assuming that his company was interested in doing that. He was in for a surprise.

Over the coming frantic months, the security team identified over a thousand documents that had been recently stolen from the company. But so many other confidential documents had been stolen over a longer time that some of the original records were now only available in tape backups. The damage was so vast—information on new products, bid status on active sales efforts—that Shields was directed to stop pulling older records of what had been taken. The mad race moved into triage mode.

What's the most sensitive information that has been stolen? Where are we most compromised? The purloined information covered a wide range—technical explanations, sales plans, product budgeting, and more. But three things stood out for Shields.

First, this was information that only a trailing competitor would be interested in. Companies like Nokia or Lucent wouldn't care about Nortel's plans for a product that was no better than their own version, at least not interested enough to risk hacking a competitor's information systems. The information taken was not consistent with someone perpetrating a financial crime, looking to raid the treasury or extort ransom. No, this was information that would only be useful to an industry competitor playing catch-up, an also-ran who was trying to quickly put themselves at technical parity with the leading players without having to pay to do the development themselves. *Good, Fast, or Cheap.*

Second, the attacks were being conducted at a state-sponsored level. There were only a few possible suspects who fit this description. And then, reviewing IP addresses of the servers used in the hack, Shields found what he feared: Most of the documents were ending up at a server registered to a Chinese shell company called Shanghai Faxian Corp. Shanghai was the location of the CCP's notorious Advanced Persistent Threat 1 hacker team.

Nortel, Shields believed, had been hacked—to its very core—by someone operating out of China. And, according to Shields, whoever it was had access to the computer and emails of the company's board of directors, as well as

Admiral Bill Owens, who had just been appointed to lead the company. If true, this presented the possibility of not only stealing confidential corporate information but of compromising the leadership of the company.

A MOLE?

The company set about cleaning and securing the system, but after years of countermeasures, Shields remained convinced that remediation efforts performed by the company were "pathetic," as he put it, and the hack was not over. By 2008, four years after the initial attack, he believed the attackers were targeting new CEO Mike Zafirovski's computer. Shields put in a request to conduct a more thorough investigation using more sophisticated memory forensics that could determine exactly what they were taking from Nortel's systems and where the information was going. The software required to complete the process cost $2,000.

Shields's request was denied.

The lack of response by Nortel was so egregious that Canadian Security Intelligence Service officials speculated about a more sinister explanation for the failure to root out the infiltrators. Michel Juneau-Katsuya, a counter-intelligence and counter-terrorism officer with the CSIS, told the *National Post* that his agency's warnings were all but ignored.

"To this day," he said, "I believe there may have been one or more agents of influence controlled by the Chinese in [Nortel] which succeeded in neutralizing our warning."

His concerns should have been alleviated after Nortel went bankrupt and vacated the building. Unfortunately, the next tenant was one of the few entities in the country big enough to need such a space: Canada's Department of National Defence. Incredibly, the DND had chosen a building that the country's own intelligence agency knew to be compromised as its new headquarters. As the agency prepared to move eight thousand military and civilian employees into the building, the *Ottawa Citizen* reported that they encountered an office complex filled with listening devices. Added in anticipation of the new high-security tenant? After a series of half denials and "no comments," Vice Admiral Mark Norman assured that *Citizen* that "it was all legacy, old-school stuff associated with the previous occupant."

THE GIFT THAT KEEPS GIVING

Shields shares a story that took place after he had left Nortel, when he applied for a job handling network security for a large American technology company that competed with Chinese manufacturers. He was called for an interview with the company's director of security, who asked him about one part of his résumé: "What's this work you describe, identifying a remote attack on Nortel's computers?"

Shields explained the hack and how it had resulted in a thorough compromise of the company's computers—one that couldn't be identified, let alone resolved through the use of standard anti-malware. He described how a foreign infiltrator had installed itself on the computers throughout Nortel, redirecting information back to servers in China.

The color drained from his interviewer's face. "Those computers—those are the ones we bought from Nortel's Enterprise division when they went bankrupt," he said. "Hundreds of them. No one ever told us." Then he paused.

"We're using them now."

Part IV

TOO MUCH AT STAKE

24

Not If It's War

Surge forward, killing as you go, to blaze us a trail of blood.
—Ren Zhengfei, founder and CEO of Huawei

In 2019, as the importance of 5G and the Internet of Things became clear and countries started waking up to the threat from Huawei, many considered it too late to do anything. Huawei was too big, too good, too cheap to exclude from the world's networks. It would cost individual companies billions of dollars to rip out existing Huawei gear and convert to other vendors and, more importantly, there *were* no other vendors, at least none that could roll out a 5G network as quickly and for the same price.

Nokia and Ericsson were coming to be seen almost as stalking horses, there to fulfill a long-understood need for carriers to retain some vendor diversity, to give 20 or 30 percent of a network to another vendor to hedge against monopoly pricing and political or company risk. Bitter irony that the global giants, who once grudgingly accepted access to China's huge market with a 20 percent cap on market share, were now relegated to fighting each other for that much share on their *own* turf.

But what, exactly, was the rationale to block Huawei from selling its products? Was it because they were being subsidized unfairly? Because they were displacing local companies? Because there was a trade deficit with China? Mike Munger, the Duke professor, had explained his view that if countries want to subsidize their exports, we should let them do it and be glad. What's wrong with another government taking money from its own citizens and transferring it to our citizens and consumers in the form of subsidized low prices? But he had come to a new understanding of the situation.

TAKING CASUALTIES

"For the first two years of the Trump presidency I thought these people were idiots," says Munger, referring to announcements of US tariffs and the threat of a full-scale trade war to counter the trade deficit with China. He still has profound disagreements with many of the policies Trump put in place, but he realizes that there is a fallacy in the context under which we consider trade with China. The argument put forward by nearly all classically trained economists assumes that companies and countries interact in a system of rivalry and competition, engaging in vigorous efforts to secure the best deals for themselves but with an understanding that they are participating in long-term trade relationships. As a result, successful global trade means that trading partners become dependent on each other. This brings about peaceful relations and creates wealth through specialization of labor.

"The problem is that I didn't understand that there was a second view," says Munger, "which is that the world is not primarily liberal and cooperative." He illustrates his point in the starkest terms. When General Eisenhower had the troops gathered for the invasion of Europe in June of 1944, he understood what to expect if he proceeded: The allies would suffer a massive loss of lives and materiel. Thousands of productive citizens would be killed. Expensive ships would be sunk, tanks destroyed; no matter how well the exchange went with the enemy, the endeavor would create a net loss for the allies. With this knowledge, why would the general proceed with a voluntary exchange that would leave his country poorer for it?

Because this scenario was not a voluntary exchange between trading partners. Those trade rules don't always apply.

Not if it's war.

In a liberal worldview, power and wealth are measured in absolute terms. Every country wants to get wealthier; it works to get the best deal possible, but if a cross-border transaction achieves that, and it is satisfied with the terms of the deal, it doesn't worry about whether its trading partner is getting *even wealthier.*

"In war, power is *relative,*" says Munger. "We will often do things that harm ourselves so long as it harms our enemies more. When a general decides whether to take a hill, he knows he may lose men, but if he expects the enemy to suffer more—if he thinks he'll impose ten times the casualties on the enemy and win the hill—it may be the right decision."

Munger sees this as the alternative way to think of our relations with China: "If we go for a trade war with China, if we restrict our imports from China, of course it harms our consumers. The president's team was not confused about that. The point was to harm China *more.*" And he makes his point crystal clear: "Because the US and China are rivals for *power,* not partners in economic growth."

This viewpoint is not one that is explained in the language of trade or economics. Global trade relies on ideas of comparative advantage, voluntary exchange, outsourcing activities that others are better at and insourcing the things that you do best. It's how all parties are lifted and become cooperating elements in a global ecosystem of trade.

But in war, wealth and power are not absolute, they are relative. It may be necessary to end beneficial outsourcing and pay more for an inferior system if this action denies your enemy wealth. It may be necessary to avoid becoming dependent on someone who views you as an enemy not a trade rival. Munger continues: "If the United States has a slightly weaker economy, and China has a *much* weaker economy, and is less able to project its power abroad, that's a benefit to the United States."

How does Munger see China advocating for its interests? "Huawei's a great example," he says. "'This is just 5G technology! How could it not be great? It benefits consumers!' Well, no. For two reasons, we should not be using Huawei: first, it benefits China by giving them a lot of revenue that will allow them to increase its global influence even further, and second, it makes us more dependent on a country that we should recognize we are at war with."

With this understanding, the free-trade economists' arguments against the US administration's combative policies carry no more water than a plea to Eisenhower to cancel the invasion of Normandy to save money.

Does Huawei reject this shift in attitude from the United States and much of the rest of the world? Is the idea of "war" excessive for a company that just makes radios? Apparently not. Huawei has responded in kind, with Ren acknowledging to his employees that "the company has entered a state of war," according to a transcript of a February 2020 speech cited by the *Wall Street Journal*. This is mild compared to comments the paper also attributed to Ren at his company's research center in Hangzhou: "Surge forward," two executives confirm he told his employees, "killing as you go, to blaze us a trail of blood."

Yes. This is war.

TRIGGERED BY COVID

As fraught as this debate already was, the global pandemic brought on by COVID-19 proved a catalyst in the world's view of China. In early 2020, after the first reports emerged of a virus infecting people in Wuhan, infections began to ravage northern Italy, then the UK, and then the US. Over the course of the next several months, as the world economy crashed and millions of people succumbed to the disease, attention turned to precisely what role China played in the crisis.

The most widely cited reports indicated that the illness started spontaneously in a "wet market," where live, wild animals were being sold. Fringe theories circulated that the disease was germ warfare, engineered by Chinese scientists in their Wuhan lab. In March of 2021, Dr. Robert Redfield, the former director of the Centers for Disease Control and Prevention, told CNN, "If I was to guess, this virus started transmitting somewhere in September, October [of 2019] in Wuhan," where he believes it came "from a laboratory . . . escaped . . ." While the true source of the virus remains unconfirmed, multiple investigations found that, as the *Financial Times* concluded in October of 2020, "The Chinese government and the WHO . . . downplayed growing concerns about whether the disease could be transmitted readily between humans." The lack of transparency led to a delay in action and may have caused millions of needless deaths.

China's behavior in the crisis, and the government's reaction to criticism, highlighted the risks arising from China's critical role in global healthcare supply chains. The world realized that crucial materials needed to fight the pandemic were sourced primarily or exclusively from Chinese factories, shining a light on poor supply-chain diversity. China, perhaps not unreasonably, was believed by some to be retaining promised exports of medical supplies, like personal protective equipment, for its own use. The consequences were grave; this was not like relying on Mexico for 90 percent of your avocado supply, where a local blight could cause a guacamole shortage. These materials were essential to preventing mass casualties across the world. Deliberately or not, supplies weren't being sent in the quantities needed.

There were more signs that the leadership of the CCP was going to take a belligerent stance to opposition. When Australian prime minister Scott Morrison called for an investigation into the origins of the coronavirus, China responded by imposing tariffs and other restrictions on goods from Australia. This was a petulant move that violated standing agreements and raised concerns over Chairman Xi Jinping's willingness to use trade power to penalize countries that crossed him. The pressure that China applied to Australia was great, with one Chinese government agency posting what turned out to be a faked image of an Australian soldier holding a knife to the throat of a young child. This provocation led Prime Minister Morrison to assert the importance of "Australia being able to conduct itself in accordance with its own sovereignty" in a November 2020 address.

As the pandemic grew and crippled the world's economy, countries reliant on good-faith interactions with China began to pay much closer attention to how China was using its trade power in general, and its technology leadership in particular. The pieces began to come together, telling a story about a China that was using its mobile technology capabilities both to subjugate its own population and to extend power across developed and developing countries alike.

China's Use of Tech to Oppress

> Winston kept his back to the telescreen. It was safer,
> though, as he well knew, even a back can be revealing.
> —George Orwell, *1984*

While George Orwell's *1984* imagined a society in which the government hid microphones and cameras in remote parks and bedrooms, China takes advantage of the fact that nearly all citizens already carry microphones in their pockets, along with cameras, location trackers, and transmitters. The Soviet Union relied on informants overhearing and filing reports on anti-state behavior, but Chinese authorities have eliminated the middleman. They have started to require "face unlock" to register a smartphone, which means their world-leading facial recognition technology tells authorities who is using a phone, and their artificial intelligence capabilities are able to process where people go, what they read, who they communicate or meet with, and what they say to them.

For those who are concerned about the dangers of China's growing influence throughout the world and wondering what its end goal is, the best proxy of China's intent toward other countries' citizens may come from considering what the CCP is already doing to its own people.

ABOMINATION

Nearly a million Uighurs, Muslim Chinese citizens from the northwest provinces, have been relocated, many loaded aboard railcars and shipped off to "reeducation camps," according to the latest reports from human rights group Amnesty International, and other credible evidence smuggled out of China. For many of them, living under video surveillance is a constant, whether in their own mosques or the barracks at their new "home."

People in free countries sometimes chafe under the ubiquitous municipal video cameras that capture everything. But what would they think of a surveillance system that could recognize a member of a targeted race or minority group and alert the police that they were in a neighborhood where they didn't belong? Most would call that an abomination.

Huawei calls it a feature.

A study released in late 2020 by IPVM, an independent video surveillance research company, reported that they had found a document on Huawei's own public website, marked "Confidential," that describes a successful test of such a capability. The document, which has since been deleted from the website but was preserved by IPVM, names a Huawei partner: Megvii, a company recently sanctioned by the US government for being "implicated in human rights violations and abuses in China's campaign targeting Uighurs and other predominantly Muslim ethnic minorities."

This is not a passing allegation of harassment or unfair treatment of a group that is considered troublesome to authorities. The US State Department, under President Trump, labeled the treatment of the Uighurs a "genocide." If Beijing was hoping a change in US leadership would bring relief from such views, the Biden administration wasted no time in affirming this position. (Unsurprisingly, China has objected to this characterization in the strongest terms.) And the role of the technology is not incidental, whether in the cameras capturing the images or the wireless networks delivering them.

Huawei responded to IPVM's discovery of the "Uighur alarm" by declaring that the ethnicity-detector feature was "simply a test" (advice to companies—don't do tests like this) and described the report as "purely slander." They told CNBC that using "modern tech products and big data to improve social management" is a general practice around the world and that their measures "are not targeting any particular ethnic groups." Later denials were softer and seemed to recognize the legitimacy of the evidence, with a Huawei spokesman saying they "take the allegations . . . very seriously and are investigating the issues . . ."

Huawei remains eager to distance itself, at least in the press, from these activities, and its partner in the surveillance service, state-owned camera maker Hikvision, hired a connected American legislator to help make their case. Barbara Boxer, former Senate Ethics Committee chair, was brought on as a registered lobbyist in January of 2021, perhaps to clarify that her client was merely assisting in forced mass internment, not engaging in complicity to commit genocide. She has since deregistered under public pressure.

RENEGING ON A FIFTY-YEAR PLEDGE

The CCP has imposed brutal policies even beyond its mainland borders, especially with the crackdown on personal liberty in Hong Kong, a former colony and dependent territory of the United Kingdom that, since 1997, has been a special administrative region of China. This crackdown explicitly violates the treaty that led to the territory's handover, in which China was legally obligated to honor certain political freedoms in Hong Kong.

One particular change in the administration of Hong Kong involved China's plans to reverse anti-extradition laws that Hong Kong had put in place in the final days of British rule, in 1997. China's efforts would have allowed Hong Kong residents accused of crimes to be extradited to mainland China, where there were few protections of rights and little respect for due process. After several Hong Kong booksellers accused of distributing books critical of the CCP disappeared and then turned up in Chinese courtrooms, citizens of Hong Kong rose up and resisted the attempts to pass laws that would make this process easier and more common.

Protests became commonplace in Hong Kong, but they drew immediate police response, with remarkable speed and accuracy. Arrests, harassment, and beatings followed. But how were the authorities identifying these rallies and protests so quickly? Brave young citizens didn't need an intelligence agency to figure it out.

In October of 2019, *Wired* magazine reported on devices mounted on streetlamps, purportedly used to measure air quality and manage traffic, but suspected of playing a much more sinister role in tracking and recording the activities of the young protesters. Those protesters sought to dismantle some of the dozens of remote wireless sensors. The images of crowds of people carrying umbrellas on rainless days show one of the simple countermeasures taken to prevent facial recognition applications from succeeding.

But the systems didn't need to see a face to know when a crowd had assembled. Whether the Chinese had placed "spoof cell tower" devices, to capture the phone numbers of people in the area, or had hacked the social media platforms being used, the police were able to converge almost immediately on any significant gatherings of protesters. This technology, like so much of the Smart Cities equipment deployed and supported by Huawei, serves as another example of how an authoritarian state might use wireless tools to identify gatherings of people, assess what they have in common, and dispatch armed officers to break them up or take them in.

CRUSHING A DISNEY SINGER

Outside of its own borders, China has been able to use its far-flung technology presence to assist other countries in cracking down on dissent. Huawei's work on the African continent goes beyond designing and deploying mobile networks and typically includes supporting the customer, whether it's the central government or a private network operator, in its every need. Sometimes that need is ensuring public safety, and sometimes that need is helping government officials maintain a grip on power, even if it means using communications technology to—reportedly—quash political opponents or find and eliminate popular opposition to the ruling regimes.

That's the claim made in a series of investigations carried out in 2018 and 2019 by the *Wall Street Journal* (and disputed by Huawei leadership and

CHINA'S USE OF TECH TO OPRESS

Chinese authorities), which show just how capable China may be at exporting the surveillance state. The danger may not lie in what Huawei can do *to* a country, but what they can do *for* it, as the world's number one exporting nation is accused of using Huawei as a tool to export its own brand of government surveillance and control.

The initial premise of the deployments is legitimate. The leaders of governments across Africa seek to achieve domestic peace and prosperity by deploying wireless networks, getting the benefits that come to every country that deploys them. But they also pursue stability by using those same wireless systems to identify, track, monitor, and even seize those who oppose the government. Few would oppose such covert surveillance when the target is a terrorist plotting against the country. But when it is a pop singer and Disney musician?

Robert Kyagulanyi grew up in a slum outside Uganda's capital city of Kampala. He's better known as the singer and actor Bobi Wine, who frequently includes socially conscious messages about poverty, public sanitation, and healthcare in his performances. (His name and music became better known in the West when, in early 2020, he released a song as a public service announcement urging people to wash their hands and be aware of the dangers of the coronavirus.) Wine would hardly seem to be a threat to Ugandan president Yoweri Museveni, who has ruled the country for more than thirty years. But Wine advocated for better health standards and won a seat in the Parliament of Uganda, where he continued to speak out against the ruling government. These actions led to him being arrested by the Museveni government and charged with incitement to violence. According to the *Times* (London), he appeared to have been beaten prior to his court hearing.

An old joke, updated, tells about an American arguing with a Chinese man about freedoms in their respective countries. The American boasts, "In my country, you can stand on any street corner and denounce President Biden as a buffoon and an imbecile, and the government won't do a thing about it." The Chinese man responds, "We have just as much freedom as you do. In my country, too, you can stand on a street corner and denounce

President Biden as a buffoon and an imbecile!" Not so in Uganda, where Wine, upon being released for the incitement charge, was later rearrested and charged with intending to "annoy, alarm, or ridicule" the President.

Throughout 2018, Wine irritated President Museveni with a series of concerts in which he spoke out against certain government policies. In December of that year, he arranged to have opposition leaders attend one of these events and rally the crowd for change. According to the *Wall Street Journal*, "A senior police commander relayed a presidential order to access Mr. Wine's encrypted written and spoken communications, including those using WhatsApp and Skype," to learn details of the event. The *Journal* says that after the police failed initial attempts, they asked for help from Huawei, and that a team of Huawei's technicians cracked Wine's communications in two days. Ugandan state security forces were able to intercept and arrest numerous attendees and organizers, many before they even reached the venue.

"The deal with Huawei is a survivor strategy to consolidate power," Wine said of the Museveni regime. "It's an all-out assault."

The *Wall Street Journal* reports that Wine's subsequent attempts to organize rallies had been foiled and his family was also now compromised by surveillance. Undeterred, Wine announced in July of 2019 that he would run for president of Uganda in the 2021 election.

HIGH-TECH IMPERIALISM

Allegations of similar incidents in other African countries have also emerged. The same *Wall Street Journal* investigation cited "senior security officials" in Zambia as the source of a report that Huawei technicians helped security services access the phones and Facebook pages of opposition bloggers who were running a news site critical of President Edgar Lungu. The sources say the information was used to track the location of the bloggers and direct a police team to find and arrest them. In that case, the official spokesman for the government acknowledged the use of Huawei technicians to aid ZICTA, the telecom regulator for Zambia, in combatting opposition news sites. The

Zambian spokesman told reporters that Huawei worked with the country's "Cybercrime Crack Squad" to intercept communications from legitimate targets like criminals, as well as to surveil opposition groups, activists, and members of the country's own media. Again, Huawei claims that they did not play any role in these activities.

These municipal systems get to the heart of the promise—and threat—of 5G networks. Such solutions can transform the way governments keep their citizens safe, with ubiquitous coverage enabled by wireless networks that move broadband video around the globe and tie it into sophisticated analytics, from facial recognition and license-plate readers to location tracking and pattern detection. They can also serve as tools of oppression, smothering dissent and isolating "troublemakers" whose only crime was failing to toe the ruling government's line.

When a senior Ethiopian official dismissed the discovery of bugs and back doors in the Chinese-built African Union headquarters, he commented that unlike some Western countries, at least China hadn't colonized the continent. But is this a new form of high-tech imperialism? Why else would Huawei employees engage in this kind of behavior?

The revenue associated with such activities doesn't seem relevant to a $120 billion company. Could technicians who assist local government teams be rogue actors, working without their employer's knowledge? Or is this part of a broader political initiative supported by China's largest technology firm? Actions taken elsewhere in Africa and around the world undercut the idea that these are one-off "rogue" operations. Financially, it's inconceivable that Huawei, after a business case analysis, considers this minuscule service revenue worth the political risk.

Or is this corporate-sanctioned, deliberate policy? That's the allegation in a federal indictment that led to the 2019 arrest of Meng Wanzhou, the CFO of Huawei, in Vancouver, Canada. The daughter of Ren Zhengfei by his first wife (Meng Jun, a woman he describes simply as "very tough"), Meng Wanzhou was arrested by Canadian authorities at the request of the US Department of Justice. They wanted to extradite and prosecute her for activities involved with breaching trade sanctions against Iran. The complex charges involve claims that she presented misleading information to HSBC (the Hongkong and Shanghai Banking Corporation) about Huawei's relation with a company delivering Huawei gear to Iran.

Whether this company, Skycom Tech, was an independent actor or a front for Huawei, the actions give further support to former FBI section chief John Lenkart's argument, that Huawei is acting primarily as an agent of the CCP. He sees deals like Skycom's alleged sale to Iran—as with ZTE, which was found to have sold gear to North Korea—as demonstrating a clear dual motivation.

"From a business perspective," he says, "what is the value of the Iran or North Korea market compared to the US market? What company would review that risk assessment and then decide to make the sale?" The companies are sophisticated global operators. They have hundreds of lawyers, risk management teams, country experts, government relations advisors.

Decisions to sell to North Korea or Iran are deliberate, the result of extensive corporate decision-making, according to detailed Department of Justice indictments. Lenkart argues that the risk of exclusion from multibillion-dollar markets like the US in order to close a tiny sale to a rogue regime is confirmation about who's pulling the strings for Chinese equipment makers. He has reviewed ZTE corporate documents signed by company lawyers and executives right up to the chairman, detailing the potential risk but directing the company to proceed.

"That's why ZTE pled guilty and paid a billion-dollar fine," says Lenkart. "They were caught dead to rights." The behavior of these companies supports the argument that China sees its involvement with other countries' governments as a means to project and impose upon the world its own philosophy on governance and control of society through the use of advanced surveillance and tracking technology.

Is there something wrong with a country becoming involved in training or even indoctrinating government or security forces in another country? It's something that the United States and other countries have been accused of doing for a long time. If one believes that there is no difference between one country and another, one can argue that it's bigotry or bias to complain when China extends its philosophy to other less wealthy countries.

All countries, even those who place the highest value on freedom and individual liberty, are presented with hard choices about how to use technology to improve safety and security without compromising personal liberty. It's a subject of constant debate and legislation across the free world.

But there is a bright line between using a license-plate reader to find a car that holds a kidnapped child and using that plate reader to intercept an opposition leader on his way to a rally.

No Way Out?

There's no question that information from Huawei routers has
ultimately ended up in hands that would appear to be the state.
—Eric Schmidt, former CEO of Google

As concerns grew about overreliance on China in global supply chains,
countries that had waved off the United States' diplomatic efforts to
ban Huawei began to reconsider. For the first time, many of them saw that
China was not playing by the same rules. They realized that China was
capable of reneging on treaty obligations or of withholding vital committed
supplies because of a perceived "insult." China was increasingly scrutinized
for creating a threat to the security or sovereignty of nations around the
world. A backlash began to build against China, raising the possibility of a
reversal on the deployment of future Huawei systems or even leaving existing systems in place.

Yet despite increased pressure from the Trump administration, or perhaps because of it, spring and early summer of 2020 passed with little indi-
cation that Huawei's grip on the market for 5G solutions would be broken.
Countries across Europe reaffirmed their relationships with Huawei or
punted on the question of a ban. In the United Kingdom, British Telecom

warned that it would cost billions of pounds and take years to completely remove the Huawei gear from the network, and even then the process would result in service blackouts for customers. Beyond concerns with communications networks, some countries feared Chinese retaliation against a ban.

WHOSE SIDE ARE YOU ON?

China's willingness to impose trade punishments on countries threatening a ban was made explicit, as was the extent of the power it had amassed though paid allies, when an open letter was published in Britain's *Telegraph*, in March of 2020. The stern letter stated that a ban on Huawei's 5G gear would not just be costly to the UK's mobile operators but would "prejudice trade relationships with China," a clear warning of across-the-board economic retaliation should Parliament restrict sales of Huawei equipment. But the letter was not from China's Ministry of Foreign Affairs.

The author was Sir Mike Rake, the former chairman of British Telecom and president of the Confederation of British Industry, the UK's largest business advocacy group.

Sir Mike seemed all too happy to put these threats—and they came across as *threats*, not warnings—in his own voice. They were oddly broad, coming from someone who claimed to speak for just Huawei, not the entire country of China. It might have made more sense if the threats came from a representative of the Ministry of Trade, with China defending Huawei as a national interest. But Sir Mike was ostensibly representing a private company, free from the grip of the CCP, distinct from the Chinese government. On what basis did he threaten the UK about China's intended retaliation?

The fact that a British leader who had made his career representing British business interests was willing to speak on behalf of a communist regime showed just how effectively Huawei had secured allegiance from its well-paid spokespeople. The fact that Sir Mike was already a wealthy man made his willingness to advocate for another country, against his own, even odder.

Odd or not, China's intimidating tactics were effective across Europe, as parliaments and regulatory groups avoided taking a hard line against

Huawei's technology. By early summer it appeared the US push had failed, and numerous articles and books predicted a Chinese victory.

And then the US made a move that nobody expected.

PULLING OUT ALL STOPS

In his 1990 book *The Prize*, author Daniel Yergin told the story of the fight to control the world's oil supplies, with "mastery of oil" being the most important tool for achieving power, growth, and sovereignty. But Yergin forecasted that computer chips might become the new "prize," the new tool that could choke off an opponent's growth. It appears he may have been right.

Telecom gear relies on microchips to operate, and many companies around the world are in the business of making them. Trying to restrict the sale of these chips to China is difficult if not impossible, with too many avenues to cheat a ban and too many companies willing to step up with their own products.

But most of the companies who make chips, at least the high-end kinds used in Huawei equipment, don't actually *make* anything at all. They *design* the chips—write the software and architect the chip—but leave it to companies called fabricators, or "fabs," to turn those designs into real manufactured silicon and gallium chips. There are only a few companies in the world that do this, or at least that make the highest-quality chips needed to build a 5G cellular network. While two major companies in the United States do fabricate high-end chips, the biggest supplier in the world is TSMC, a Taiwanese company whose output is vital to Huawei.

And this is how the US was able to make an extraordinary move: While the American manufacture of chips isn't essential to Huawei's success, America's supply of the equipment used to *make* chips is. Without machines from American companies, fab plants around the world, including TSMC, can't make the more advanced chips needed for 5G equipment.

Having failed to apply diplomatic muscle effectively, the United States used its trade and industrial strength to invoke a "nuclear option." In May of 2020, the US announced to fabricators around the world that anyone selling chips to Huawei would be denied these vital machines.

As a group, they folded.

One company after another announced that they could not operate without American equipment in their fab plants. And suddenly the question facing Europe, Australia, India, and the rest of the world wasn't whether they would continue giving contracts for Huawei networks, but whether Huawei could deliver on the networks they were awarded. Indeed, without access to chips from any of the non-Chinese fabricators, Huawei's ability to compete effectively in any market became a question. Huawei acknowledged that their ability to build equipment without the needed chips was threatened, as their stockpile dwindled. At the company's annual conference in September of 2020, Huawei's chairman Guo Ping put it bluntly: "Right now, survival is the goal."

Couldn't China build its own fab plants and make its own chips? Yes and no. The country is on a crash program to develop such a capability, but the money required is staggering, and even money can't buy the skills needed. Each state-of-the-art fab plant costs ten to twenty billion dollars to construct. And even these plants require know-how that isn't simply purchased. The ability to manufacture microchips is reliant on having a team of experts working together on a wide range of complex activities. China can't just write a check and have a factory built and staffed.

Not for lack of trying, though: China claims to have hired 10 percent of Taiwan's chip engineers, three thousand so far, luring them with enormous salaries to build capability on the mainland. This program is off to a slow start, however; in 2019, China's most advanced process was a 28 nm planar technology, something Taiwan had been making for ten years.

THE GLOVES COME OFF

Faced with this crisis, China turned the screws tighter on European countries, using trade pressure and threats in an effort to compel them to keep their markets open to Huawei. The results have been mixed.

Among the stronger rejections of Huawei was that of the United Kingdom, which, in July of 2020, announced one of the strictest bans on procuring new 5G equipment. UK officials went further, requiring that service providers remove all Huawei gear already in their networks—including 2G, 3G, and 4G gear—by the end of 2027, costs and blackouts be damned.

Sweden was next, banning all Huawei gear from its upcoming 5G auction, which brought a direct threat from China to penalize national champion Ericsson, whose sales in China make up an important part of its own revenues. Ericsson, restricted as they may be from securing a high market share in China, still knows that nearly 10 percent of its sales come from the China mobile equipment market, while less than 1 percent come from selling into their own country.

In a case of strange bedfellows, the CEO of Ericsson, Börje Ekholm, implored the Swedish minister for foreign trade to reconsider the ban, reflecting the pressure his Chinese customers were able to bring to bear on him. In text messages obtained by Sweden's leading newspaper, *Dagens Nyheter*, Ekholm implored the trade minister to talk with the country's telecom regulator, even reportedly speculating that he might be forced to take Ericsson out of Sweden. The country held to its decision.

Finland, home to Europe's other national champion, Nokia, saw China's aggressive response to Sweden's announcement and chose not to ban Huawei, insisting that Finland did not take a stand on a product's country of origin. However, Finland did pass legislation requiring network operators to deactivate equipment that is a "risk to national security," effectively putting a stop to further Huawei sales—for now.

France began withholding approval for carriers to deploy Huawei gear in September of 2020, citing "European sovereignty," as opposed to a flat ban, and leaving the window open to later approvals.

What of Germany, the economic and industrial engine of Europe, which had been one of the countries to continue supporting Huawei deployments? Chancellor Angela Merkel, leading the country for fifteen years, faced down the first-ever rebellion by her own Christian-Democratic–led bloc in February of 2020, as she refused to ban Huawei.

Why would she dig in so aggressively? By far the largest company in Germany is Volkswagen, at $283 billion in 2019 revenues. How important is the China market to Germany's most important company?

It's not just that China is Volkswagen's largest market. VW sales in China are greater than in every other market in the world combined. VW sells twice as many cars in China as it sells in all of Europe, including Germany, and China was not shy about threatening the loss of the market, with the Chinese ambassador specifically promising retaliation against the German auto industry. "There will be consequences," warned Ambassador Wu Ken on a video shown on the *Handelsblatt* website, before slyly raising and "dismissing" the possibility of China banning all German cars—28 million sold in China in 2019—on safety concerns. "No," he continued, providing little comfort. "That is pure protectionism."

The pressure caused pause, but the security risk that Huawei gear presented to Germany was deemed to be too great, and in September of 2020, Merkel agreed to support a new law imposing strict security requirements on all "high-risk" vendors selling into Europe's largest market.

Some European countries have simply issued long-term contracts to non-Chinese vendors, claiming the bid process had naturally led to them, but not explicitly banning Huawei. Others are considering allowing Huawei gear to remain in their networks, including some that are planning to deploy it into the edge of the 5G networks.

The United States, Australia, Japan, Taiwan, and New Zealand have explicitly banned Huawei from the countries' networks.

The cost of refusing Huawei will be high. Replacing existing equipment is expected to increase deployment expenditures by billions of dollars; additional billions will be spent on price premiums that Nokia and Ericsson charge over Huawei's rock-bottom prices. These efforts mean that China's presence in Europe's communications networks, including 5G, may be reduced or eliminated entirely. But is it really worth it?

27

5G Is Different

5G will be the backbone of the digital economy, and
the digital economy is increasingly the backbone of the
world economy. It's important to get this right early.
—Rush Doshi, director of the Brookings Institution

The winners and losers of the 5G war won't just be the vendors or operators of network hardware. *Everyone* will be affected. Depending on how the solutions are rolled out, and who does the deployments, consumers will see changes to their quality of life, whether in entertainment, transportation, medical care, or elsewhere. Governments in cities across the world will be able to provide better security, power management, and traffic control, or be hamstrung in their efforts. Military readiness and battlefield effectiveness, too, will depend on whether the armed forces can take advantage of the networked battlefield or whether their communications are ineffective, insecure, and compromised.

It may seem extreme to say that the prosperity and security of countries, companies, and individuals around the world are at stake, but as with every industrial revolution that preceded it, those who reap the advantages that the Internet of Things brings will move ahead of the pack.

The characteristics that make a business or government effective will be greatly enabled by 5G communications, the first wireless technology that is purpose-built to do more than just connect people to other people or information. With earlier generations, the bulk of mobile network activity consisted of people calling or messaging other people or accessing the internet. In the world enabled by 5G, mobile communications will more often take human hands off the device and make the connections invisible to the people who are benefiting from the service. Devices will talk to devices. Information will talk to information.

TWITCHY

The technological advance in 5G isn't really about speed: 5G solutions significantly improve bandwidth, increasing speed from around 10 megabits per second to 100 mbps or more, but adding bandwidth is relatively easy in wireless. It's like adding lanes to a highway. It may not be cheap, but it's not complicated and we know how to do it. To a great extent, 5G's increased speed is delivered by simply using more spectrum than 4G uses. No magic there.

Improving latency—reducing it—is a bit trickier, and that's an improvement that opens a world of possibilities. Latency, the time it takes for what you do on your end to make the round-trip to the destination and back, matters more and more. The original 2G cellular data connections of the 1990s had latency approaching 1,000 milliseconds; it took nearly one second for your command to reach the destination and return your information. Not a problem when your desire was to download an email. That latency won't do if a factory wants to scan a defective product whizzing by on an assembly line and flick it into the trash bin. For that application, or for virtual reality simulations or remote surgery, you need a "twitchy" service—one that gives a response in a hundredth of a second or less.

Each succeeding generation of wireless has improved on this, and 5G is the twitchiest of all, with the possibility of latency in the single digits—as low as 2 milliseconds—but achieving this creates problems that don't just affect the service provider.

Latency is added with each mile your command needs to travel to reach the server or whatever is at the other end. It's added each time the command passes through servers on the way to its intended target, and it is further added if those servers are overloaded. Engineering out the latency of 5G networks is one of the great challenges, and the urgency around it is leading to changes in how 5G networks are being designed.

One way to help achieve this will be to push the processing closer to the end user, which will have a profound impact on security: by putting the source of information or computer processing closer to the end user's cell tower, all the way at the edge of the network, the service provider eliminates potentially thousands of miles of travel and multiple servers along the path, significantly slashing latency. Still, this creates opportunities for mischief—or worse—from an untrusted vendor of that radio gear at the edge. Regulators don't seem to have understood this yet, but this edge/core distinction is important enough to explain in better detail. An analogy helps:

The CIA does not give tours of its Langley, Virginia, headquarters. It's just not done. Anyone can sign up to climb the Washington Monument or tour the White House, but almost no one gets into the core of the Central Intelligence Agency's operations.

For those lucky visitors who do get to enter the complex, the visit begins weeks in advance with the delivery of background information that lets the security office vet them and approve their pass. On the day of the appointment, visitors are met at the entrance by heavily armed security officers. After confirming that they have been pre-approved, the visitor advances to the next level of gates and scrutiny. And so on.

On a recent visit, as one lucky guest was walking up (finally) to the main doorway of the iconic building, he pointed to a high fence surrounding the close-in parking lot. "Does that fence have some kind of sensors built in to set off an alarm if an intruder touches it?" he asked.

His host turned to him with a serious look on his face. "No one would ever get that far."

But many people have met with their CIA partners at the multiple satellite offices in the area. No need to deliver social security numbers weeks in advance. No gates, no armed guards. Just pull up next to the bagel shop in Reston, Virginia, and walk right in. (Of course, the receptionist behind the desk may be packing a Heckler & Koch sidearm . . .) Meetings are regularly and easily held in any of the CIA's offices around the region, or even the family-style restaurants in the McLean, Virginia, area.

The difference is that the CIA ensures that there isn't any significant cost if they are compromised at the edges of their infrastructure. The valued assets—files, equipment, laptops—are kept at the core, protected physically from intrusion, compromise, interruption. They don't leave the secure core, and you can't get to them.

Owners of wireless networks have long shared a similar perspective. The core of the network needs to be protected; it holds the servers, the billing engines, the databases. It controls and allocates calls and data sessions, ensures continuity of service. The edge of the network just consists of dumb antennas and radios. They *connect* to the core, but they are only vulnerable to the extent that disrupting them can disrupt service to the specific region they cover.

At least that's how networks used to be. As new 5G networks are being engineered, architects are redesigning them to deliver capabilities like the low latency needed to enable medical procedures. This means placing servers and software right at the base of the towers or even at company locations. That profound change in network topology slashes the distance information needs to travel. Your inquiry no longer needs to make the round trip from the tower to the remote server thousands of miles away; it can be processed and returned to you directly from the tower.

The downside is that it creates a significant new vulnerability in 5G networks. That formerly dumb tower may now host the secure corporate or personal data needed to deliver the mobile service. It may house the classified software that enables the army's war game to be executed. It's as if the

CIA started allowing its analysts to take their files and laptops out of the agency headquarters and work on them at the local bagel shop.

In addition to pushing resources out into the network, the nature of the communications changes between the remote tower and the core. What used to be a simple voice conversation between them is increasingly a complex data and signaling session. As a result, if that tower is trusted by the core, any breach at the edge lets the bad guy through the palace gates. The implications are great for companies that consider letting untrusted vendors play *any* role in their network. It's becoming clear that even the edge—the dumb cell tower—isn't so dumb anymore.

Dangerous Designs:
Why We Need to Care

If Huawei is allowed to deploy their equipment
into our 5G networks, they won't need a back
door; they can use the front door.
—A senior intelligence advisor to the
president of the United States

With the new architecture of 5G, the job of securing the network against enemies grows exponentially harder. The idea of digging a moat, building a firewall around the precious core of the network, is insufficient. Now, every part of the network, right out to the edge, is a potential point of vulnerability for the entire network. Even past the edge, 5G networks create a far larger "attack surface" than anything we've had before. With the Internet of Things, not only is each device a threat, but there are far more of them, and devices will be cheaper and simpler, and less managed, than the laptops and smartphones that used to log in to the network. Operators, more than ever, need to be able to trust the vendors of network elements. And even then, operators must be careful.

SOMETHING FISHY

One recent example that made the rounds in the cybersecurity trade press illustrated the risks of IoT deployments and stunned the normally "seen it all" cybersecurity community.

A clever tech geek charged with maintaining the aquariums throughout his employer's Las Vegas casino figured he had found a way to save money and time. He deployed off-the-shelf remote sensors to the massive aquariums situated throughout the gambling floor and connected them to the internet through Wi-Fi. This allowed him to log on from any PC and remotely monitor water temperature and oxygen levels in the tanks, adjusting the aquatic environment from his desktop as needed. A harmless initiative, so removed from the world-class security measures required by the casino's IT experts that he didn't even submit the change to the top brass for approval. By the time the casino was done paying a cybersecurity company to figure out what hit them, the casino's high-roller customer database had been compromised and the IT manager was contemplating a future repairing slot machines.

Experts explained how the casino was hacked. The remote sensors and controls are typically shipped with default passwords. Given the supposed low stakes, most operators don't bother resetting them, and this case was no different. What potential risk could there be—that a hacker would turn the temperature up and cook the koi? But an outside agent had bigger fish to fry.

A hacker was able to access the unsecured fish tank sensors via the internet and, through them, enter the casino's operations database where he proceeded to access and download all the personal and business information on the casino's high rollers. It didn't take a George Clooney–led team to stage an *Ocean's Eleven* raid on the casino, just an unsecure tank bubbler that was connected to the company's otherwise secure core information systems.

WORSE THAN A NUCLEAR ATTACK

If a financial loss to a Vegas casino doesn't elicit tears, Justin Fier, a former intel operative now with Darktrace, a cybersecurity company, shares examples that serve as a stark reminder of the vulnerability of wireless networks and connected devices. Why are they so vulnerable? "Software tends to trust other software more than it trusts people," Fier says. "By putting wireless devices into the network, a trusted path gets created.

"The IoT has been willfully insecure from day one," he continues, "and now we're talking about rolling out a new technology that is both exciting and horrifying at the same time."

Fier had a customer whose global restaurant locations serve billions of meals a year. While doing an audit for the company, he discovered that newly deployed temperature sensors in the refrigeration units were connected to the internet—and unsecured. "I could have launched a Stuxnet-like attack on the refrigerators units, raising the temperatures well above safe levels and making them look like everything was operating as normal," he says, referring to the successful 2010 attack on Iran's nuclear centrifuges that reassured operators that the machines were working normally as they spun apart. It wouldn't be until people started falling ill or dying of salmonella poisoning that the authorities would have started looking for the culprit. "The client had no idea these systems were even online and visible. They were just refrigerator controls; no one thought to secure them."

The list of remote-connected industrial sensors is already long and expected to grow dramatically as 5G takes off. But it's not the only danger that can be exploited.

The COVID crisis spurred a frantic rush to take advantage of other remote wireless equipment, whether forehead temperature scanners mounted at the entrance to offices or always-on cameras set up in class-rooms for remote students. The deployments were made with a tie-back-the-safety-valves urgency, and the potential dangers are just now being contemplated as details of our lives are being moved from the real world to the cloud, with the wireless world of 5G serving as the natural delivery vehi-cle once the devices are set up.

It's not just civilian sites that are vulnerable. Military and intelligence facilities using advanced security measures find themselves vulnerable to the hacking of wirelessly deployed remote equipment.

"We caught a biometric fingerprint scanner being compromised," Fier recounts. This was a device used to scan people's fingerprints and grant them access to secure areas. "All of a sudden it had internet connectivity," he says, "instead of just talking to its own control system's devices. We watched everything the intruder did. We saw the intruder download the database that contained usernames and access levels from the company's human resources server through the fingerprint scanner, then—this is the interesting part—they uploaded a new database. And the size of the file was different. Information had been added." Fier and his colleagues then watched the intruders clean up after themselves. That means they had the opportunity to add names to the secure list or change the places a given person had access to, and no one at the company would have known the database had been altered. "My client was in complete shock."

One of the key reasons for poor security, even with otherwise security-conscious users, is the sheer number of devices in the 5G Internet of Things world. When Darktrace plugs into a client's network to map out the devices, Fier says, "Nine times out of ten, they'll tell us something like, 'You'll see ten thousand devices connected,' and we'll show them fourteen thousand."

The billions of devices that will be connected through 5G each present an opportunity for a breach of the entire network, a huge attack surface. These are not major network elements managed by operations centers at the network provider. They aren't even smartphones, with PIN codes and assigned chips. They are all the way out past the edge of the network—boxes costing a few dollars each. Millions of them today, and billions of them soon.

And they are made in China.

Fier describes the dangers: "We've seen intruders enter a university's research database through an air-conditioner unit that was networked so the building engineer could monitor and control it. The building with the HVAC unit was still under construction when it was hacked," providing access to a major research university's confidential data, including research that was proprietary.

"The ICS (industrial control systems) space scares me more than anything," he says. "What would be worse: a nuclear attack or an ICS attack where the entire East Coast loses power?" He argues that a long-term loss of all electricity could impose a greater loss of life and financial damage than a nuclear bomb going off in a city.

SHUT DOWN

Fier's warning doesn't seem far off, as millions of Americans recently experienced the devastating consequences of a major power crisis. In February of 2021, a record-breaking winter storm hit the United States, with Texas bearing the brunt of it. A period of extremely cold weather—the coldest in more than seventy years—sent power demand through the roof as residents consumed record amounts of electricity and natural gas to heat their homes and businesses. At the same time, temperatures plunging to below zero froze natural gas pipes and knocked wind turbines off line.

The combination of high demand and restricted supply pushed the Texas power grid to the brink of complete failure, and on the morning of Monday, February 15, the people running the Electric Reliability Council of Texas (ERCOT), which operates the state's electrical grid, decided to initiate rotating blackouts to avoid a total collapse of the power system. In theory, this proactive measure means that people should have only experienced controlled, short-term outages, typically limited to less than forty-five minutes. Instead, millions of Texans were plunged into darkness in subfreezing homes, some for more than three days.

By the end of the crisis, the storm cost dozens of lives and over $100 billion in damages and losses. What's alarming is how close the state was to even more disastrous outcomes. Bill Magness, president of ERCOT, told the *Texas Tribune* that the grid "was seconds or minutes [from possible failure] given the amount of generation that was coming off the system." The results of such a collapse would have meant that the state could have lost power for *months*, according to Magness.

In those panicked moments, ERCOT leadership had to make instant decisions based on updates on their system's generation, distribution, and demand for power. Increasingly, such information is delivered through remote terminal units that capture and report critical data on power grid elements and even allow remote operation of controls at those sites. These units are often connected through wireless links and increasingly will rely on 5G networks to protect the grid. In a crisis, managers cannot take the time to dispatch crews to remote locations, and without reliable information from these sensors guiding their actions, equipment can catch fire and substations can be destroyed.

There were no reports of reliability problems with the remote terminal units in the Texas grid. But if an enemy were to hack or disable these types of remote units during a crisis, there is no telling how much damage could be done to a power grid. In the moment of decision, operators require complete confidence in the reliability and integrity of the control elements. Without that, there is no way to avoid catastrophe.

And it's not just the public infrastructure that is coming to rely on wireless sensors and reporting. The 5G-reliant factories that are now being put into service use hundreds or thousands of remote wireless sensors to monitor activity, track progress, and adjust processes. Tied into automation systems, these factories can have a significant performance edge over operations using older, more manual techniques. But they create new risks.

A pharmaceutical factory may have remote sensors that report when a process has moved outside of acceptable levels. Automated responses can then shut the production line until the problem is fixed, preventing wasted time and materials and avoiding the manufacture of substandard medicines. But if someone were to spoof a false report from those sensors, the factory— or multiple factories—could be taken offline, stopping production of vital medicines during a critical health emergency. Discovering and ending the hack could be difficult, and adjusting operations to bypass the sensors and bring the factory back on line might take weeks, or longer.

Could China hack critical infrastructure or a factory control system in the United States? Could it bypass the security measures installed by whoever built the control system? It might not have to, if it built that system and still has its hands on the daily maintenance and operations. With China's dominance in all things wireless, is it too dangerous to let China's companies build these IoT networks?

THE FIRST
HACK

The idea of hacking a wireless communication is not new—it's older than the earliest wireless networks. Old enough that the first victim of a corporate sponsored wireless hack was named Guglielmo Marconi, known as the "Inventor of Radio." He was both a technology pioneer and an ambitious businessman, and in 1903 he was eager to show that messages could be sent securely, an important consideration as the wired telegraph was becoming widely embraced as a tool of commerce, diplomacy, and military communications.

A crowd was assembled in the lecture theatre of the Royal Institution in London, to receive a message from Marconi, located 300 miles away at a station in Cornwall. Marconi had boasted, "I can tune my instruments so that no other instrument that is not similarly tuned can tap my messages."

The claim may have been technically true, but another person had apparently "similarly tuned" his own instrument, and just before the demonstration was scheduled to begin, the telegraph in the theater began clacking. The dots and dashes that represent the word "Rats," appeared over and over again in Morse code. The telegraph in the theater was unfortunately connected to a Morse printer, which translated the dots and dashes into words and projected the output onto a large screen for all to see. After a pause, the message became more pointed. "There was a young fellow of Italy/Who diddled the public quite prettily," it began. The rest of the message was sufficiently shocking that the papers of the day did not print the details except to mention that it included allegations of Marconi taking liberties with the hacker's wife.

Days later, the culprit revealed himself. He was Nevil Maskelyne, a music hall magician who claimed he was undertaking the hack to warn the public that the technology was not as secure as advertised.

The real motivation was more mundane and commercial. The Eastern Telegraph Company had invested massive amounts of capital in undersea cables to carry wired telegraph signals to all corners of the world, and this new disruptive technology presented a risk of damaging their hold on the market, so the company hired Maskelyne to spy on Marconi and reverse engineer his "secure transmission" technology. His flamboyant prank succeeded in making sure that this weakness in the wireless telegraph became widely known.

29

If It's So Risky . . .

If we do this right, it could render our smartphones the least
interesting thing about the future of wireless technology.
—Jessica Rosenworcel, chairwoman of the FCC

If the danger is so great, why even build this 5G wireless Internet of Things?
Because the benefits of this interconnected world are enormous.
It promises to deliver a fourth industrial revolution, with broad improvements in health, poverty reduction, physical safety, and general economic
growth. The challenge will be to achieve this extraordinary value in a way
that doesn't leave users vulnerable to interference, or worse, from those
with ill intent, whether internal suppliers or outside attackers. Detecting
bad activity won't be easy, because these next-generation wireless networks
are based on connecting things to other things, information to information.
The magic will happen in the background, where it may be harder than ever
to even detect problems. Consider this example of the difference between 5G
and what has come before it:

In the early days of cars, people needed to use a key to unlock their door
and turn on the ignition. Then, they simply needed to have a key fob in their

pocket, and the car would unlock when they got close. With the advent of 4G apps and modern cars like Tesla, people didn't even need a key anymore; they can unlock and start their car with an app on their phone.

With 5G, people won't need to use their phones at all. Wireless networks will notify a car that its owner just paid their lunch tab or passed through the security check on the way out of the office. The car will consult weather data and decide if it should start itself to preheat or cool the vehicle. GPS data will notify the car when the owner is near, and it may even come to pick them up, using sensors, cameras, and cloud-based artificial intelligence. None of this will require the user to take actions on the 5G network; the multiple remote sensors and transceivers will invisibly do this all on their own, using machine learning to adjust practices to better satisfy the customer. This isn't science fiction; many of these solutions have already moved from lab to prototype to early commercialization.

LONG DISTANCE

One of the most anticipated capabilities of 5G is the advent of telemedicine. Millions of people got their first taste of a "light" version of this during the early COVID lockdowns, meeting with doctors via videoconferences. The convenience was compelling, even if the experience wasn't always satisfying. But these Zoom health visits pale in comparison to what ubiquitous wireless networks are beginning to enable.

Companies have already rolled out enhanced tools that allow doctors to diagnose and treat patients more thoroughly via remote wireless connections. Soon, 5G wireless solutions won't just deliver high-definition video images and complex telemetry on the patient's vital signs but will enable doctors to administer treatment, from altering the settings on a wireless pacemaker to adjusting insulin doses from an implanted pump if a person is entering diabetic shock. Those vital remote commands, and the integrity of the connections that deliver them, will rely on the trustworthiness of the companies building and operating the networks.

There are even bigger plans in store for telemedicine. The first explo-ration of wireless-enabled surgery has begun. The basic concept of remote surgery isn't new; in 2001, a doctor in New York City was connected to a robotic system in France, where he used the local networked surgical instru-ments to remove a patient's gallbladder. The connections were carefully established over fiber optic lines, with redundant fiber lines used to ensure a more reliable link and reduce latency, which would cause a lag between the doctor's movements and the actions carried out on the patient.

These last two issues—reliable connectivity and latency—are the big hurdles that 5G finally clears, allowing doctors to attempt *truly* remote sur-gery, where there is no reliable wired connection.

Outside of the major population centers of the world, high-quality fiber optics are not ubiquitous, and they may not be fully deployed for decades. But just as early cellular telephony eliminated the need to dig ditches for copper wires, 5G may be the first, and perhaps only, high-bandwidth/low-latency connection for rural areas. For a country as large as China, and with as big a divide between the high-tech cities and the still-backward coun-tryside, the idea of remote surgery via 5G connections is compelling, and China's lead in this technology is impressive.

Still, given China's record of human rights violations, the country might have been advised to pick a better procedure to demonstrate its lead in this space.

In March of 2019, Dr. Ling Zhipei, chief of neurosurgery at the People's Liberation Army General Hospital in Hainan, performed the world's first 5G-based remote surgery. The bandwidth of the wireless connection deliv-ered a crisp high-definition video image, and the low latency ensured that the movements of Ling's hands in Hainan were immediately transmitted to the remote robot instruments in Beijing. The operation was a success, and the patient was all too happy to testify about the results. "I feel good," the unnamed patient was quoted as saying in the state-controlled *China Daily*. And the procedure?

The doctors implanted a chip into the man's brain.

To be fair, the patient was said to be suffering from Parkinson's disease. The operation, a legitimate medical procedure, but new and rarely used, consisted of implanting a deep brain stimulation chip into the head of the patient. Still, such a demonstration hardly seems like an ideal way to dispel fears of the CCP as a totalitarian organization bent on complete control of the hearts and minds of its citizens.

The patient was not asked how he felt about participating in the advancement of 5G telemedicine in such an intimate way, nor did the media get his view on having the procedure done only weeks after it was first tested on an animal. The reporter knew better than to label this a dangerous stunt, and he dutifully repeated the surgeon's explanation about why this procedure had to be done remotely, despite the fact that the doctor works in Beijing, where the patient was located: "The operation took place during my Hainan rotation. A patient with Parkinson's in Beijing needed surgery and couldn't fly to Hainan."

Of course. It was just a good use of 5G technology to help the man resolve his medical condition. Still, for a procedure that is not considered an emergency, no explanation was given as to why the surgeon didn't wait a week until he was back in Beijing. Or simply use the fiber optics links that already connected the hospitals.

The point was made: In the absence of a skilled surgeon, in the absence of a broadband wired connection, talent can be delivered anywhere you have a wireless network with the characteristics found in 5G.

DOWN TO BUSINESS

If this sounds impressive, the industrial side of the transformation may be even more dramatic. The next wave of wireless networking will allow all aspects of production to make a leap forward in efficiency, quality, and safety. More so than in day-to-day lives of citizens, and probably much sooner, the 5G revolution will transform industrial activity, and, in the breakneck pace of global corporate competition, any country that falls behind on this process will find its industries losing business and fading away. This isn't about sensors reporting information to people so they can do something. This is about sensors reporting information to algorithms that use AI to make changes and direct other systems to change, or issue commands down the supply chain that never see a human's eyes. If a human gets involved, it's a failure in the process. Factories taking advantage of this technology will make laggards obsolete. In *all* industries.

The benefits will extend beyond commercial applications. In practice, 5G can enable a world where every element of a city or public infrastructure system is linked, not just to each other but back to a central brain that commands and controls the operations all around us. Such a transformation holds the possibility of profoundly improving the quality of such vital services as power generation and distribution, drinking water, public transportation, and other services provided by the government.

For example, the benefits of intelligent remote connectivity for a modern power grid, which may include millions of miles of power lines or gas pipes, are enormous. Sensors may be deployed at distant locations to deliver a simple reading—perhaps a temperature or pressure figure—back to a control panel, which enables managers to know if there is a problem. An intelligent 5G network can enable the problem to be resolved—say by opening a valve or closing an electrical circuit—without needing to send a technician to a distant location. Not only is this cheaper than a "truck roll," which can cost hundreds of dollars in the city and thousands elsewhere, but it enables more thorough monitoring and management than would ever be attempted using wired assets.

This creates a far more valuable service even as it opens a greater opportunity for bad actors to gather data on people, businesses, government officials, and military operators around the world. With the new capabilities, it won't just be possible to hear what people are *saying*; an intruder will be able to know—and *affect*—what they do: 5G is no longer a gateway for outside enemies to listen to what we're saying; they will be able to interfere with the most basic activities of life, from riding in our driverless cars to entering our secure homes to running our factories or farms.

Yet, given the core technologies that make up this next-generation ecosystem—wireless networks, remote sensors and actuators, mobile handsets—the network vendor expected to lead the market is one that is not trusted: Huawei. The countries of the world may have a choice: lag in 5G because of a reluctance to allow Huawei in, or open the door to having a Chinese company build their next-generation networks. It should be clear why we need a third option.

Part V

TAKING IT BACK

Building a Response

When disruptive innovations have to fight with other
innovations for resources, they tend to lose out.
—Clayton Christensen, American consultant and author

While it may be necessary to ban untrusted suppliers of network gear, that is hardly a solution. With China offering the most compelling end-to-end 5G solution, our challenge is not to find a way to just get by without the best technology or the most capable vendor. The challenge is to develop our own solution that allows free countries to deploy the world's best communications networks so that businesses, government institutions, and citizens can quickly and securely obtain the benefits of 5G for themselves without the dangers of relying on China's champion company. The countries that win this race will reap the rewards that have historically gone to those who embrace each industrial revolution, with gains in both absolute economic growth and relative political/military strength.

The danger remains that, if we can't develop a competitive and compelling 5G solution ourselves, China will be able to deploy superior systems and move ahead of the rest of the world in industrial and agricultural productivity, scientific research, and military sophistication. The potential impact is that great. Whether Chinese-made 5G systems are deployed around the world or not, the rest of the world will be dealing with a rising China whose intentions may be a threat, and we need to chart a path that doesn't allow them to surpass the capabilities of the rest of the world.

The new course requires changes in governmental policy and industrial activities, but this is not a call for central planning or government control of any sector. While the potential answer may touch on structured industrial policy, the heart of the solution must lie with delivering superior commercial approaches.

How has China won the commercial battle thus far? It played to its strengths, but more importantly, it played to our weaknesses. The West, especially the United States, has always had a more freewheeling competitive culture. Free-market companies are driven by their need to deliver returns to shareholders, and that can lead to a shortsighted, siloed way of thinking. Frank D'Amelio, the former CFO of Lucent who was named CFO of Pfizer in 2007, sums up why the West ended up in their current bind: "Everybody was doing what was best for their company. No one was thinking about what's best for the country. That was the conundrum." But this isn't a simple problem to resolve; you can't wave a wand and cause companies to act in the interest of anyone besides their owners. It's the heart of a free-market, capitalist system. And while these characteristics present the opportunity for China to play one faction off another, they also hint at our advantage.

Though China has its own chaotic entrepreneurial energy, which has contributed greatly to its success, its companies' interactions with trading partners have been ultimately governed, or at least constrained, by their CCP masters and directed toward specific national missions and goals. This has been true both of the sclerotic state-owned enterprises, which act completely at the direction of the Party, and the more independent, entrepreneurial companies like Huawei, which pursue profit and growth where they can find it, but are guided, corrected, and supported through incentives, aid, and implied penalties, with the understanding that, ultimately, no institution in China has the leeway to oppose the CCP's desires. And not only is their path rigidly set, they make no secret about what it is. As former FBI section chief John Lenkart says, "The neat thing about China is that they're very open about their geopolitical goals. They're good communists, they have five-year plans. The great thing about those plans is they *tell* you what they're going to do, then they actually *do* it."

We can use this to our advantage.

LOCKED IN AT THE TOP

Huawei reached unmatched size in part through relentless execution and massive government support, operating in a sector where success has been determined by scale. As 5G has emerged, the incumbency afforded by their past wins has fortified their position and provided great inertia; once a service provider purchases their gear, the cost to switch vendors becomes nearly prohibitive.

Scale may not be the only way to win this game. Historically, scale has derived either from the company's size itself or from the imposition of a technology standard that requires all suppliers to comply with a strict set of characteristics. For example, by naming GSM as a wireless standard for Europe in 1989, the benefits of scale were enjoyed by all who built to that standard because of the huge supplier ecosystem that emerged.

That leaves the West a few options to meet Huawei's scale:

- Create a giant of similar or greater size to go against them
- Adopt an industry standard that excludes Huawei
- Pursue a radically different path that changes the game and creates scale in a new way (or introduces another determinant of success)

The first option would likely require beefing up competing vendors. The United States spent trillions to prop up the economy after COVID; spending

$10 billion to juice Ericsson or Nokia would be trivial, and it would transform their ability to compete, wouldn't it?

Maybe not.

If a pile of money from a US government initiative suddenly lands on the conference-room table in Stockholm, only a fool would expect the budget and development process to retain any discipline. It would be payday at the marine base, with projects that should be killed launching nonetheless with all the discipline one would expect when you tell a Swedish engineer to be frugal because "this is Uncle Sam's money, and he doesn't want it wasted."

It's too late for this kind of a move, anyway. A strengthened Ericsson or Nokia, even a *combined*, strengthened Ericsson and Nokia, as some have called for, would still be a tiny force compared to Huawei. The European vendors had a combined revenue of around $50 billion in 2020, less than half that of Huawei, and if scale is the determinant of success, there is little value in taking on Huawei by slightly lessening their scale advantage.

The second option is not so easy. We could create a split standard for 5G, with an exclusive standard that bars China from participating in networks built around the world. Such an approach virtually guarantees that no one anywhere will enjoy economies of scale, and that customers would suffer under the complexity of having multiple incompatible standards in place. Considering the direction technology is moving, with more open, interchangeable elements, such a schism would be a step backward, and a hard one to ever remedy.

The third option is more open-ended. How exactly would one "change the game"? To the extent that any transformational solution will require a national policy—and this can't be solved without government engagement that spans and coordinates among many activities in many countries—government-led programs rarely deliver rapid, efficient, innovative solutions. History has shown that such programs tend to devolve into corporate welfare programs and winner picking, with the winners invariably being the largest, most well-established players or the most politically connected. They don't change the game so much as lock in the old rules.

LET THE GOVERNMENT FIX IT?

America has long been considered the world leader in microchips, but the US doesn't really make them. Companies like Qualcomm, the world leader in mobile chips, design them, but they must be manufactured by a company with a factory, a fab plant. This is where the US was getting its butt kicked by the Japanese, and this is what SEMAT-ECH (Semiconductor Manufacturing Technology), an industry consortium formed by the US government in the mid-1980s, was intended to reverse. Funded with a then significant half-billion dollars from the Department of Defense and DARPA, SEMATECH worked to shore up American chip manufacturers who were being crushed by better, cheaper Japanese microchips. Companies throughout the industry assigned engineers and managers to a program that worked for years to regain the lead from the Japanese. The result?

"Honestly, this was a massive failure," says Steve Papa, CEO of Parallel Wireless, an American start-up offering innovative solutions to carriers deploying any wireless standard. "The result of SEMATECH is TSMC (the Taiwanese chip giant). They are the ones who gained." He argues that the structure of such organizations leads to domination by the largest, most entrenched companies who can afford to staff and co-fund them, and that leads to a bias toward the types of policies that are least likely to react nimbly to advances in technology. America's largest chipmakers were helped to the detriment of smaller innovators with less scale, and the largest American companies were then choked out by larger international vendors. In effect, we ensured that the "Innovators' Dilemma," (as described in Clayton Christensen's seminal book on the challenges that large, established companies have in transitioning to new ways of competing) kept the sector from jettisoning current technology approaches even when opportunities for discontinuous, radical improvement emerged.

Some disagree, arguing that SEMATECH saved the US industry and reversed the gains of the Japanese. In the early nineties that seemed to be the case. But the proof of the pudding is in the eating, and within ten or fifteen years of SEMATECH reversing America's losses to the Japanese, the industry had faltered again. Today, at the high end, only two major American chip fabricators, Intel and GlobalFoundries, remain dominant.

No, the answer lies in the other direction—to find a way to unleash the creative and productive potential of an entire society. It's been done before.

Our Superpower

We cannot regulate the future with yesterday's means.
—Jack Ma, Chinese business magnate

To regain the lead from Huawei, the role of government should not be to dictate the technology road map and compel industry acquiescence. The first objective must be to clear a path for a uniquely American approach to innovation, one that reverses the game and draws on our core nature, the same characteristics China has used to exploit us so far. The US has fundamental qualities, drawn from its culture, history, and resources, both intrinsic and brought in from immigration of creative, ambitious minds. These qualities can be leveraged to turn the game on China.

Sun Tzu called for stealth, cunning, surprise to defeat a greater foe. In the Western tradition, the Israelites did not meet Goliath by sending out their largest warrior; they won by meeting the threat asymmetrically. Entrepreneur David got the job done after the state failed to come up with its own winning plan—and he didn't play by Goliath's rules, either. Our solution must likewise play to our relative strengths. And we have one in particular.

PERMISSIONLESS INNOVATION

In the last twenty years, the United States has produced a string of companies that are unmatched by the rest of the world. The companies have emerged as game-changing leaders, inventing new markets and creating massive value in short periods of time. Companies like Uber, Airbnb, Amazon, Google, and Facebook have sprung up, not with the careful aid of government assistance, but despite the best efforts of government to regulate them, tax them, stop them. What do they have in common with each other? These businesses were created in frequent, often direct, conflict with the rules.

They violated the laws:

Private homes are not zoned as hotels.
Retail stores must charge a sales tax.
Regular people aren't licensed to drive a taxi.

And these businesses frequently operated in conflict with social norms:

No one wants to get into a car with a stranger.
One does not research a blind date's social history.
Why would the world want to see a picture of your avocado toast?

So far, the transformational business successes of the twenty-first century are nearly all American and they all engaged in *permissionless innovation*. These companies were not gently encouraged by legislatures and regulators to launch and grow. The companies *fought* those institutions every step of the way. They sidestepped regulators. They broke industry models. They turned the old-fashioned ecosystems on their heads. Often, they broke the law and challenged, in court, the government's right to enforce it.

Adam Thierer, a senior research fellow at George Mason University and author of *Permissionless Innovation: The Continuing Case for Comprehensive Technological Freedom*, says he didn't coin the term, crediting it to Admiral Grace Hopper, the computing pioneer who famously attributed her breakthrough innovations for the US Navy to "asking forgiveness not permission." But Thierer has been a student and advocate of it in his books and lectures through the Mercatus Center at George Mason University. A self-described former telecom wonk, he also credits China with granting its own innovators leeway to create new approaches "within certain boundaries," which is both a function of its efforts to emulate American models and a surviving remnant of China's own pre-revolutionary spirit of creativity and invention.

"There are more innovators in China than in Europe," Thierer says, "but they also have state values that can't be messed with. You don't rock the boat dissenting with the Communist Party." In other words, the CCP has implicit or explicit limits on what it will tolerate from its entrepreneurs, and when it decides the answer is to dominate 5G through the lowest priced, broadest selection of gear, compliant with the industry standard, it doesn't want anyone shifting investment to some iconoclastic new bet against the house. Established companies and entrepreneurs in China know that if they cross a line, the penalties may be severe. And there is no recourse.

Herein lies the answer to how the free countries of the world can avoid mortgaging their future to China's technology hegemony. By relying on permissionless innovation we can take China on, not with something it *can't* do, but with something it *won't* do. Could the CCP release people to act as they choose, challenging the government and thumbing their nose at the system? Of course.

Would they?

Not a chance.

This approach is consistent with our strengths and limitations. Huawei is already operating at far greater scale, with more R&D dollars than all their major competitors combined; there is no technology investment program that can outrun them head-on.

But though Huawei has greater scale than any competitor in the world, they don't have more resources and ingenuity than *every* competitor in the world. That is, even Huawei can't match the power of a permissionless ecosystem based on open access for all comers, a solution that allows an equal seat at the table for every start-up, every Cal Tech junior working out of their dorm room, every industrial giant looking to diversify into a new market.

Such disruptive activity is the engine that has historically led to technological breakthroughs and economic growth. Permissionless innovation is the process—it's too unstructured to call it a process—it's the *condition* where entrepreneurs can innovate, break rules, create risks and dangers, and pursue discontinuous solutions, all without having a government or industry regulator looking over their shoulder telling them to stop. Permissionless innovation can't be proctored, sponsored, approved, or managed by a third-party authority. And it is dangerous: investments get wiped out; companies get crushed; vital, venerable institutions are rendered obsolete, all without a distinct intention to do so.

You don't find this as a line item in a five-year plan.

Nothing could be harder to support at the government level. Politicians and regulators don't secure donations and lock up votes by supporting someone they don't know who is going to make something that may be illegal and sell it to unidentified consumers who haven't asked for it. It's hard to establish a political constituency around that kind of approach, which is why there are rarely advocates for it in the government despite its demonstrated contribution to economic growth and prosperity. And if the payoff to elected leaders is slim in democracies, you can be sure that the embrace of iconoclastic initiatives is no ticket to the CCP's inner circle.

The amazing thing is that it ever happens, given how easy it is to quash. In India, which sends so many brilliant technologists to thrive in the US, those same people have not been able to achieve similar results in a system that makes money by withholding permission and charging admission. In India, you can't do what you want to do unless you pay the right bureaucracy for a license, check all the boxes, and spend so much time waiting in line at the permit window that the market window closes.

Yet such behavior is how some of America's greatest innovations have come about. So many of the creators of our markets started as rebels, or worse, and not just in technology. For example, Herb Kelleher challenged the tightly regulated airline industry when he formed discount carrier Southwest Airlines, and had to take his case to the Supreme Court to confirm his right to fly Texans at low prices. Permissionless innovation is a solution based on liberty and the freedom to challenge the norms of business and society. This plays to our strengths. More importantly, it plays to China's weaknesses.

The Chinese system that Huawei must operate in is *fundamentally unable* to abide such forces of anarchy and mayhem, even though this model has led to trillions of dollars of value creation in world technology markets just in the past decade. As a product of permissionless innovation, Facebook alone has created more market value than all such companies in Europe *combined*. It's not that China can't replicate such innovation. *It won't.* Culturally, politically, China would never allow this uncontrolled, subversive behavior to define its markets.

That is not to say China cannot innovate; this old belief has been dispensed with, as companies like Alibaba, Baidu, and Tencent invent popular and successful solutions. Still, Chinese authorities have little patience for activities that directly threaten the government's control over society. Chinese executives do not seek to provoke the government and face off against them in their country's corrupt and opaque courts. Entrepreneurs know there is no equal treatment under the law, and business executives aren't safe from persecution, no matter how wealthy and connected they are. Innovators are aware that there is a limit to the ability to change the game and push the envelope. China's most successful business leaders live their lives at the whim of the CCP, where loss of freedom is always just one misstep away.

NO RECOURSE

Tim Cook, Apple's CEO, showed his ability to thumb his nose at the feds, refusing to grant FBI agents access to an accused terrorist's iPhone in 2016. He knew that the worst consequence might be a court case, a fine, or regulatory action against the company.

Not so for Chinese billionaire and real-estate tycoon Ren Zhiqiang, after he made comments in a published essay that were considered insulting to Chairman Xi, leading to his abrupt disappearance in 2020. Although Ren, the son of a former vice minister of commerce, was politically connected and himself a senior member of the CCP, he quickly learned that no one is immune to the wrath of the party leadership. Months after he disappeared, Ren resurfaced in police custody and was brought to trial, where he was accused of a series of crimes ranging from misappropriation of state funds to bribery and abuse of power. Complex as the charges were, the closed trial was completed in a day, and he was found guilty and received his sentence. Anyone else who has considered speaking up learned that the punishment for being disrespectful to Chairman Xi is eighteen years in prison. To be fair, Ren referred to Xi as "a clown stripped naked who insisted on being called emperor." But if that were the criteria in the US, the Trump presidency would have seen most of Silicon Valley taken into custody.

Little cover is afforded to executives who step out of line, even if they have brought glory to China and showed that the country can match the best America gives. Jack Ma, the richest man in China and the billionaire founder of Alibaba, China's answer to Amazon, vanished in the fall of 2020. Immediately, there was speculation that he was in state custody. His crime? He had given a speech accusing the country's regulators of stifling innovation. His views came on the heels of comments to the contrary by Xi's close ally, Vice President Wang Qishan, who called for an emphasis on regulation and the primacy of the state over business.

Ma's response "was about risk-taking, putting your neck on the line and not minding the instability that comes from that," commented George Magnus, an associate at the China Centre at Oxford University. Such talk is "anathema to the philosophy of Xi Jinping's party." Indeed, days after his comments, Ma was called to meet with regulators, purportedly to discuss his latest bold move: the spin-off and IPO of Ant, his online digital finance operation that was at the cutting edge of technology and innovation. The IPO—reportedly eight hundred times oversubscribed—was cancelled by China's regulators. Investigations were ordered into parent company Alibaba on charges of monopolistic practices, and Ant Group was told to back off its growth plans. Although Ma has since resurfaced, and his expansion plans are back on track, his tone has been more subdued. The point has been made.

The Ant venture shows that China's entrepreneurs are capable of great creativity and innovation, led by native executives who blaze trails not yet cut by Western leaders. As such, one would be hard-pressed to find a more perfect example of how China deals with the prospect of such permission-less innovation. The brilliant, beloved, wealthy, and powerful entrepreneur Ma faced the destruction of his businesses and the loss of his own personal freedom for proposing a less tethered approach to innovation. If China is willing to inflict such damage on its own economy and market leaders in response to such approaches, who would be foolish enough to propose the next disruptive idea?

"You know that the threat gets worse as you go down," comments Marty Cooper, the father of the handheld mobile phone. "What about the guys who have careers ahead of them, who are afraid to do anything because if they fail they are going to get smashed down?" It's an unfavorable environment for innovation, to say the least, compared to the American business culture during the time of his own breakthrough work in cellular invention. "I don't know how I ever survived twenty-nine years at Motorola other than they were just so tolerant of my obnoxiousness, of the fact that I did have failures."

China built its system on terrifying its own populace about the fear of crossing a line that is not explicitly marked. The values imposed and enforced by the CCP are inconsistent with the idea of turning the populace loose to do their thing. It's not that the entrepreneurs in China *can't* respond with their own freewheeling, anarchic approach. They *won't*.

A PLACE TO RUN

There is no magic in the water in the United States. There is no blessing that is bestowed on people when they are born here, or when they take the oath of citizenship or receive their green card. The term "American exceptional-ism" is fraught; it's perceived by many as a jingoistic view of America—and Americans—as being better than others, blessed and destined for a higher purpose. But if the country *does* have a claim to exceptional innovation and creativity, it derives from the combination of a freer marketplace and a workforce of the *world's* best and brightest, coming to this country to flex their ideas. In America, more than any other single country, one gets a meeting of minds from across the world, people who came here specifically to pursue their vision with a belief they could succeed and reap the rewards in an environment where the laws, courts, and social norms are more inviting to the outsider and permit greater freedom.

Is this still true? If so, it isn't permanent, and the business environment is not as free as it has been at other times in our history. But today more than half of America's biggest companies, the Fortune 500, were founded by immigrants, or children of immigrants, bringing fresh thinking, diverse approaches, perspectives that both build on and enrich the traditional American approach.

While there are also companies in Europe run by nonnatives, very little of this cross-cultural fertilization takes place in China. The country's culture and laws make it hard for a non-Chinese national to found and lead a successful technology company, and as of 2021, all of China's major companies were run by people born and raised in the country or of Chinese heritage. China is not looking to deviate from the present cultural values. Curiously, almost no technology companies outside of China are run by people born and raised in China. For all their exports, management style at the CEO level is not one of them, and the unwillingness to stray from the party line may be a major reason.

Permissionless innovation is the closest market equivalent to the concept of exceptionalism that can be found, and the ecosystem that would support and enable this new approach can take root and thrive in other free countries, like Germany, Korea, Japan, India, Brazil. While not all of these countries have the same entrepreneurial culture as the United States, they give far more freedom for their companies to operate than China and, more importantly, they have a mutual trust between each other, a faith shared with the US that is not shared with China.

The Role of Government

One of the things we're trying to do is view the China
threat as not just a whole-of-government threat, but
a whole-of-society threat on their end. And I think it's
going to take a whole-of-society response by us.
—Christopher Wray, FBI director

China took advantage of the chaos and lack of coordination within competing nations to seize the lead in wireless innovation. That chaos
can now unleash forces of breakthrough innovation that no government-
directed and restricted system can match. China's model will become its
own worst enemy. This is where a global response to Huawei's dominance
can take root. This is facing them on *our* terms.

It's critical that governments avoid the temptation to get too involved in
helping this along. To the extent the US regulators attempt to carry companies to success, they will limit growth. Or, as Professor Munger puts it, "If
permissionless innovation is America's superpower, politics is Kryptonite."
The forces of government bureaucracy that have prevented India from
fully realizing its economic potential can equally throttle US or European

entrepreneurs, preventing them from delivering on their ability to respond to Huawei.

Permissionless innovation doesn't set up the scenario the way politicians would like, with the elected officials preening about being the one who greased the skids and made it happen. And it's not a conservative/liberal or Republican/Democrat thing. The Democrats used to argue for the assurance of regulations and oversight, and the Republicans used to push for business freedom. Now, they both just argue about which party controls businesses better. This will have to change if we are to unleash the potential of permissionless innovation.

CREATING THE ENVIRONMENT TO INNOVATE

Most of the role government can play in enabling permissionless innovation, as described so far, consists of staying out of the way, but there are active roles that will be essential in driving development of a successful 5G capability. These include:

- Asserting authority to ensure that regulators and industry groups pivot to more competitive, open standards
- Providing companies with cybersecurity insights and basic research that are beyond the budget and know-how of even the largest private-sector players
- Ensuring the protection, through law enforcement and the courts, of intellectual property
- Releasing ample wireless spectrum to allow new services to enter the market

Each of these four areas is critical to developing a successful ecosystem.

STANDARDS BODIES

If requiring permission from a government regulator to offer a new product is a time and money sink, just consider the burden of requiring permission from your *competitors*. For a disruptive new entrant, this can be prohibitive. Although new entrants often have to seek approval from a regulator who has been "captured" by the incumbent industry (see Uber facing off against taxi and limousine commissions), the greater barrier comes from incumbents using industry standards to block new entrants from entering the markets.

These industry groups play a vital role in telecommunications in general and wireless in particular, where spectrum must be shared and market viability requires interoperability between certain network elements. Any networked solution, whether railroad tracks or mobile phones, needs its standardized gauge, its agreed-upon frequency or protocol that lets a phone talk to a cell tower whether in New York or Los Angeles. These standards bodies are quasi-governmental, being made up mostly of industry representatives and academics, with participation and oversight from government regulatory bodies, and they have played a critical role in the development of wireless standards and successive generations of mobile communications. The current lead bodies, the 3GPP and International Telecommunications Union, meet to determine the rules of the sector that everyone must play by, and their decisions make or break companies and technologies.

Such standards can aid the incumbents, says BroadSoft's Mike Tessler. "The standardization of wireless 3GPP has played to Huawei's strengths," he continues. "There's not a lot of flexibility in the 'speeds and feeds' (the performance specs), so whoever can sell it for less wins all the business." What that means is that standards bodies decree a set of product metrics that every product must adhere to. By reducing the value of differentiation, they effectively make competition a matter of cost, which means scale. Huawei gets the lion's share, other big companies get leftovers, and new entrants get left out.

This is not necessarily malevolent; it's essential to agree on common protocols. The challenge lies in the tendency of markets for those protocols to evolve into a system that helps and protects the groups setting the standards. The groups must be compelled to adopt standards that don't serve to block new entrants from participating.

This would require a difficult shift in the standards organizations, and recent developments have only made it less likely. Over the past few years, the composition of the standards organizations has changed, with a growing presence of participation from China-based engineers, which has created a significant culture clash. The standards groups are made up of competitors, but the nature of the process requires more collaboration, leading to agreements that leave everyone with a seat at the table when the music stops, or at least a fighting chance for a seat. That is not to say everyone works in perfect harmony; there is a constant jockeying for position and arguing for advantage. Elbows are sharp. But the language of the standards bodies is more technical and academic; everyone gets to make their pitch based on technical and economic models, with the participants trying to find a path that makes the most logical sense for the industry as a whole. There are winners and losers, but the debate is just that—a debate, unlike the battle that takes place outside of the standards groups when these companies hit the street to sell against each other.

This approach does not sit well with the Chinese representatives, who see the process entirely as a forum to advance their position over that of their competitors, taking share and positioning themselves for more market power and influence. Reports have long circulated of multiple Chinese companies realigning their positions to settle on a standard that supports Chinese vendors. Given the fierce competition between companies like Huawei and ZTE, this can only be attributed to a greater master pulling the strings and compelling them to align around national interests.

The number of Chinese participants on the decision-making boards has surged from almost zero in recent years. As one board member of a major standards group confides, "A lot of the members don't like seeing Huawei chairing a committee, but there's no legitimate basis for us to keep them out. It's a problem." China's contention-based approach is recognized as a threat to the more objective, merit-based model, but there is little these groups have been willing to do when confronted with a company that has such significant share and is willing to invest so heavily in supporting its cause.

Given the importance of these organizations in wireless telecom, the first role of government is therefore to ensure that the standards bodies are not commandeered by the incumbents in general, and Huawei in particular. If this is impossible, given the nature of such groups, the government must clear a path for entrants to enter the market without passing through the gates placed by these groups.

CYBERSECURITY

Security is a perpetual weak spot, with companies understandably devoting their investments toward revenue opportunities, not national or even corporate security. There is, however, a nationwide network of highly skilled technologists ready to step in and deliver against this challenge.

America's Department of Energy is the parent agency for the National Laboratories, which are recognized as an expert, if underused, resource with particular knowledge in security issues. They are ready to serve as both a source of practical research and a clearinghouse for nonclassified intelligence insights, delivering support to American efforts without dictating product direction.

There's Dan Elmore, for example, the director, Critical Infrastructure Security & Resilience at the Idaho National Lab (INL). He made a career in the air force, where his responsibilities included leading presidential contingency communications and overseeing secure communications around America's nuclear defense and attack capabilities. Today, he leads INL's Wireless Security Institute, which he created in 2019 to bring advanced R&D capabilities to those working to secure 5G cellular networks.

His team focuses on the research and technical gaps that the carriers and vendors may not be able to address within the constraints of their budget, expertise, or access to unique capabilities. In the quarterly-profits-driven world of the technology sector, such endeavors get short shrift. It's hard to redirect budget away from winning new customers to preventing a cybersecurity problem that may never occur.

The difficulty with this business-only thinking is that the carriers end up with a woeful gap in their understanding of the security risks to the network and little ability to address them. As Sprint's Steve Elfman said when asked if his company's validation lab could spot a cyber threat secreted in a Huawei component, "We wouldn't know what to look for. That's NSA stuff." Or a job for the INL.

To be fair, as bad as many of these businesses are at security, the National Labs are equally so at technology transfer. The National Labs are explicitly prohibited from competing with private industry, but instead focus on solving critical technical challenges to complement or augment industry and academia. Further, "We're not great at marketing ourselves," Elmore acknowledges. "So many visitors come out and say, 'Wow—we didn't know this existed.'" That needs to change.

The most valued role for the National Labs will be developing solutions for security problems and making that output available to in-house teams at the equipment vendors and operators. The government will need to make partnerships with the Labs a more ubiquitous part of the industry, which they can do by requiring equipment makers and service providers to get a "seal of approval" confirming that security measures have been met, like the stamp required from Underwriters Laboratory, which affirms compliance with safety standards. (The Cyberspace Solarium Commission included a similar proposal in its March 2020 report, though the idea did not make it into the first wave of legislation that resulted.) In that role they could promote such simple requirements as banning factory-default passwords and adhering to benchmarks on other security metrics.

Could the INL help identify and clear security threats in network equipment?

"Absolutely," says Elmore. "It's not easy, but we've done similar things for industrial control systems and the electric grid for years in this lab." His team has expertise in finding things that don't belong. "As we get smarter, the other side gets smarter. They're going deeper down in the stack, in the software, firmware, or hardware."

It takes time to reverse engineer the suspect gear, he says, so his lab has developed models of looking for the sensitive use cases, zeroing in on critical parts of the network. This is a proficiency that equipment makers and service providers may not have; it's not cost effective for each company to develop its own cyber expertise at this level, so INL is already helping some of them secure their early 5G systems, conducting research that is sometimes used by other agencies to put out "zero day" bulletins when the lab identifies exploits. The focus is entirely on security, and the billion-dollar budgets behind America's National Labs provides resources that can't be matched by even the best equipment vendors.

The role Elmore sees INL and other offices of the National Labs playing seems well suited to helping secure the networks against bad actors and ensuring that an American-based solution eliminates the problem we're wrestling with from Huawei gear. "Look at business leaders," Elmore says. "They're motivated to make money. We're not profit motivated, we're security motivated, mission focused. The best capabilities are where the government labs have a partnership with the private manufacturers, where we dissect the 5G problem with a security view."

IP ENFORCEMENT

Creating a better solution won't mean much if it can't be secured against immediate theft from our opponents. Before any improvements can be made, we need to ensure that they won't simply end up in the hands of the people we're competing against. A 2018 report from the Office of the United States Trade Representative found that "Chinese theft of American IP currently costs between $225 billion and $600 billion annually." Law enforcement agencies and court systems need to ensure that intellectual property rights are being protected and enforced. This isn't an easy thing to fix, but one solution is simple: The US needs to become more aggressive in its policing, prosecution, and prevention of IP theft.

The challenge comes down to raising the cost of stealing intellectual property and lowering the business payoff from the theft.

Actions can include placing trade sanctions against companies caught engaging in the practice, like placing them on the "Entity List." This is a blacklist published by the US Department of Commerce that prevents other companies, including vendors and banks, from dealing with companies on the list without a federally issued license. Such a designation can make it nearly impossible for a company to engage in international trade.

SPECTRUM ALLOCATION

One critical change is simple to describe if difficult to enact. As one FCC bureau chief put it, "Spectrum plus capital equals innovation." What he meant was that you will naturally get new solutions if you make more of the needed ingredients available. Spectrum is to wireless markets what land is to agricultural markets; the more you have available, the more value you can produce. The federal government needs to continue freeing up spectrum, much of which is held for government or military use, and turn it over to companies who will make it more valuable through their development of market solutions.

There are two challenges to doing this. First, government agencies in general, and the military in particular, are stingy about giving up any spectrum they already have the rights to. They use it to keep America safe from attack, and they don't want to jeopardize this mission by squeezing their activity into a smaller slice of the airwaves, although the arguments they put forth don't always hold water.

Secondly, the main consumers of new spectrum have generally been the incumbent wireless service providers. They have a good claim to it: they provide most of the connectivity in the country today; they have most of the talent and technology, at least in sheer numbers; and they are willing and able to pay the most in FCC auctions. The 2021 spectrum auctions raised over $80 billion, with most of that coming from companies like AT&T, T-Mobile, Verizon, and the large cable companies. New entrants were, not surprisingly, limited by the bidding power of the incumbents.

Allocating all new spectrum to the companies with the biggest invest-ment in the status quo is not a recipe for breakthrough innovation, despite their essential role in delivering 5G services. Some of this spectrum must be made available under rules that prevent it all from ending up in the same hands doing the same thing.

The government can enable an environment that turns loose forces of entrepreneurism and innovation, sheltering new entrants from barriers to entry and encouraging established players to deliver solutions that are pro-ductive and secure. By doing so, they will allow the growth of a new eco-system that will unleash the technologies of the next industrial revolution.

Riding the Wave of Chaos

I've searched all the parks in all the cities
and found no statues of committees.
—G. K. Chesterton, English writer

If government authorities could be convinced to enable a freer environment, with more open standards, better IP protection, and wider ownership of spectrum, what would permissionless innovation look like? How would the different sectors take off and grow, and what would the benefits be?

"Some sectors are born free," says Adam Thierer. "Robotics, virtual reality, 3D printing; there's no government commission overseeing their actions. But some are born into captivity, with well-established regulatory institutions watching over them. Aviation, space exploration. The taxi industry. Uber and Lyft were born into captivity, and yet they broke through." Even space exploration has hit escape velocity from government control, as Elon Musk's SpaceX has turned sector economics on its head and pulled off a string of breakthroughs.

What about telecom and the 5G marketplace?

"For 5G solutions," Thierer muses, "permissionless innovation is going to be harder to make a reality."

TRANSFORMATIONAL BENEFITS

How might a successful implementation impact the competitiveness of alternative 5G solutions? It can deliver:

- Breakthrough technologies in the elements of mobile networks (the deployed equipment and chips that compose a 5G system's cell towers, phones, and other gear throughout the network), leading to cost and quality improvements
- A flood of new applications that run on those networks (turning companies like Verizon into app stores for enterprises and consumers)
- Great leaps in the flexibility and customization of individual networks in factories, hospitals, offices, farms, ports, and more (through the control and ownership of custom-built private wireless networks, which have traditionally been owned by the service providers)

For each of these, there are strong signs that new developments in technology will free innovators from the constraints that have limited them and left control in the hands of a few large companies.

NEW EQUIPMENT

The elements of wireless networks are currently supplied by a dwindling number of huge equipment makers like Nokia, Ericsson, Samsung, and Huawei. And by the standards of the industry, the architecture is generally closed and the components are non-interoperable. That means someone operating a cellular network can't swap out a Nokia radio for one made by Ericsson or plug in a Huawei switch alongside the Samsung units. More importantly, there is no breaking apart the racks of equipment in the base stations located at the bottom of cell towers, which include radio transceivers that generate the wireless signal that connects with your cell phone, and baseband units that translate those radio signals into a form that can be sent back over the fiber network.

This is not a technical detail; these boxes are the equipment that make 5G networks operate. They are what Huawei (and Nokia and Ericsson) sell, the heart of a radio-access network. Those two units are tightly coupled, and the most brilliant inventors of a new kind of radio know they have no chance of selling it into such a closed system. It would be like an entrepreneur trying to independently develop and sell a better dashboard to use in your Mercedes; the market is closed to outside innovators. Even getting a new radio licensed and approved by the FCC could break the budget and kill the timeline of an innovator.

Recent developments are starting to change this and appear poised to launch a modern wave of innovation. An approach called software defined networks (SDN) will tear down some of the biggest barriers to new, innovative entrants. SDN, and a closely related approach called network function virtualization (NFV), sounds complicated but is actually quite simple to understand.

The easiest way to explain SDN is to compare it to what happened over the past twenty years with mobile phones. Years ago, if you wanted to make a call, you had a cell phone. If you wanted to send an email, you needed a BlackBerry. To take a picture, you used a digital camera, and to record video, you needed a camcorder. Appointments, addresses, and phone numbers were stored in a personal digital assistant. Each device was a combination of hardware and software dedicated to a specific function. The result was that people had to buy many purpose-built devices, each great at what it did, but cumbersome and expensive.

When smartphones were introduced, the problem was solved with one piece of hardware that could perform any number of functions, depending on what software was loaded on the phone. Perhaps the camera wasn't as good as the best Nikon, but it was good enough for most needs. And although this new smartphone might cost more than the old phone, the cost covered all the devices previously needed, and it was a far more efficient way to do all these activities, because it had a much higher utilization rate. Before, if you weren't taking a picture, all the hardware associated with the camera was idle and useless. But now that same screen, battery, and processor might be productive sending an email. The software on the device defines the functions.

Each box in a 5G network has a specific function, and today each one is purpose-built by a company like Nokia or Huawei with special hardware and software designed to perform a specific function. One device looks up a database to confirm that a caller is authorized to make a call. A different box in the cell tower hands off a mobile call to the nearest tower as a driver moves down the road. This leads to an enormous number of unique, custom devices throughout the network, and they generally only work with other devices from the same manufacturer, which typically leads a service provider to buy everything from the same vendor. SDN allows the operator to use standard computers in place of these custom devices, with software defining the function of each computer. Importantly, this allows the service provider to *disaggregate* its network elements; each piece can work with elements from any vendor. Such a profound change eliminates the need to buy everything from one vendor. Some service providers are already deploying early SDN networks, with AT&T leading the way. Rakuten, an innovative company in Japan, is the first company to build an entire 4G network on this SDN principle, and they are upgrading it to 5G. The companies are finding it drives down complexity and increases equipment utilization rates from 10 percent to as much as 40 percent, leading to huge savings.

"A typical telecom network might have fifty thousand distinct devices in it," explains telecom industry analyst Roger Entner. "In a software-defined network there might be five to eight." In the Rakuten network, for example, there are only five different kinds of boxes that are deployed: two types of antennas, two pieces of equipment to control the antennas, and servers running on a more powerful version of the same CPU that is already used in the computers made by any of a number of manufacturers. Each one can be configured with software to do all the tasks these fifty thousand devices used to do.

Rakuten has pushed the frontier further, deploying another model that stands to transform the industry. Called Open RAN, for open radio access network, this approach provides a standard that, if adopted, will break the bond between radios and baseband units and allow anybody to create and slide in a better component alongside the other elements of a 5G system. Open RAN effectively allows the edge of the network, the cell towers and antennas, to share the benefits of SDN. This might seem like small change, a ripple in the pond, but it may turn into a tsunami for the industry.

And it does something much more important for the competitive landscape. The components of cellular networks, the hundreds of billions of dollars' worth of equipment made by Huawei and a handful of giants, can now be replaced effectively and cheaply by companies from any one of a number of suppliers in the information technology equipment industry. With SDN and Open RAN, specialized equipment that was previously made by the telecom suppliers can be replaced with off-the-shelf general-purpose servers from Dell, HP, Fujitsu, NEC, or other manufacturers. This approach may not always eliminate company scale as a factor, but it reverses the game; the biggest server makers are larger than Huawei, and by letting HP and Dell into the 5G market, they can displace proprietary gear with generic equipment, in effect, turning the scale and ingenuity of the IT world loose in the telecom equipment world.

APPLICATIONS

These equipment cost savings may pale beside even greater gains in new applications and services, as opening the network opens the floodgates to innovation. Until now, no one besides the original vendor could write applications for a public network. It would be like home viewers trying to change the scripts on NBC's Thursday-night lineup; take what you get or change the channel.

In the days of voice-only telephone networks, new applications took years to develop. Caller ID and Call Waiting were introduced by phone companies when their equipment vendors were good and ready. The huge switches had millions of lines of code written by engineers trained on that company's proprietary system; adding a new feature, even one as simple as

Caller ID, took years of development and testing, and it could only be delivered by the switch vendor's team.

With the advent of open, software-defined networks, businesses, farms, and municipal governments will be able to procure mobile services that weren't created by the cellular carriers or equipment makers, but rather by other companies, from small, entrepreneurial start-ups to giants like Google or Apple.

The benefits of this can be seen from the way internet service companies, free to innovate without permission of the network provider, have delivered far more valuable applications over their networks, from videoconferencing to social media to movies on demand. These features weren't created by Nokia or Verizon, but by innovative companies sending their applications "over the top" directly to the user.

As mobile network applications are increasingly written by new entrants or even freelancers, the services like smart cities and connected factories will be freed from the restricted menu of options offered by the handful of vendors active today. Such a move neutralizes a dominant vendor by taking away their edge in developing apps. When that happens, any new vendor of radio gear, no matter how small, can have a massive library of applications, too, perhaps more appealing than those offered by Huawei, because they are sourced from millions of developers the vendor has never met.

NETWORK OPERATIONS

The third element, network operations, may present the most radical departure. With 5G comes the possibility of customized wireless subnetworks that are owned and operated by the businesses, warehouses, farms, or cities that use them, and supported by nonnetwork carriers who specialize in industrial automation.

Today, wireless networks are run by the service providers who buy spectrum, construct networks, and manage the services they provide to consumers and businesses. A factory can hand out cell phones to its employees or deploy wireless sensors throughout the facility, but the network they run on and the services they receive come from the service provider.

Companies aren't so constrained in other areas; they can have a private company design and build their office computer network to specifications, with whatever security elements or features they desire. A customized on-premises cellular network? Not so easy.

One barrier to this is being torn down, as a new class of spectrum license is being issued, not controlled by the large service providers. Called Citizens Broadband Radio Service (CBRS), it combines the best characteristics of public wireless networks with free uncontrolled Wi-Fi spectrum. Today, only the owner of a particular slice of cellular spectrum can use it, a protection that ensures that Verizon, for example, can control the quality of service they provide over the airwaves they bought at auction from the government. Public Wi-Fi, on the other hand, can be used by anyone anywhere, as long as they follow rules about power levels, but it suffers from quality-of-service problems including interference from microwave ovens and other Wi-Fi networks that are free to operate in the same place and frequency.

CBRS offers the best of both worlds, allowing companies or cities to operate their own networks without waiting on the carriers to offer the service. Even the name chosen for the service, CBRS, is a reference to the permissionless wireless communications that were offered half a century ago, citizens band radio. CBs allowed anyone who bought a radio made by any vendor to talk to anyone else on the network, using whatever brand of radio they chose.

With these tools at their disposal, the participants have the ability to reinvent the mobile world and create services that have never been imagined.

Who, exactly, will these participants be? Will the ecosystem be made up of new entrepreneurs, or will it be dominated by the same companies who control it today? The answer holds some surprises.

34

The New Ecosystem

When one door closes, another opens; but we often
look so long and so regretfully upon the closed door
that we do not see the one which has opened for us.
—Alexander Graham Bell

What will the 5G ecosystem look like if we are able to align these capabilities—government and industry—to respond to China's dominance? What companies will emerge to lead the sector, and how will they work together in the new model? It's impossible to know for sure; some of the most promising new companies may go bankrupt or be absorbed by rivals. Companies that haven't yet secured their seed round of funding will become leaders. Too much remains uncertain to make a clear prediction, and it would be arrogant to presume otherwise—remember McKinsey? But here are some scenarios of what we may see in the coming years.

THE NATURAL COMPETITORS

The incumbent equipment suppliers like Nokia, Ericsson, and Samsung realize that their leadership days are over, at least if they continue playing the game the way they used to, trying to beat Huawei head-on. The path to future growth, or even future existence, is narrowing, but it's not closed. Already, they are turning their attention to software-defined networks and Open RAN.

Successful companies have an enormous challenge when they need to pivot from a profitable, successful business model to a less profitable one in order to survive and fight another day.

Telecom analyst Roger Entner is one of the experts who believes these companies will not only survive but thrive in this new environment. "If you're the dominant provider and you don't disrupt your own market constantly," he says, "some small disrupter will catch up with you and kill you. The writing is on the wall, and Nokia and Ericsson realize this." He believes that expertise built over the decades by the incumbent companies will give them an edge in the design and deployment of networks, even those based on newer, more open models.

THE ESTABLISHED NEW ENTRANTS

Other well-established companies have also begun allocating development resources to enter 5G markets that are adjacent to their own. In some cases, the technical challenges aren't great and the growth opportunities are significant.

On the hardware side, IT giants like Dell, HP, NEC, and Fujitsu already sell boxes that can serve as the all-purpose devices that make up most of the new networks. It will not require a profound redesign—perhaps speed improvements—to optimize their servers for a telecom network scenario and, given the size of the market opportunity, the business case is compelling. To the extent that many of these networks may be sold directly to enterprises—factories, farms, hospitals, municipal governments—companies like Dell and HP are already staffed with sales and support organizations serving these segments.

On the software side, industry giants who already sell special-purpose network software are developing broader offers to support 5G networks.

VMware, owned by Dell, is a leader in network virtualization software, as is Red Hat, owned by IBM.

With new workable hardware and software entrants, there will be a significant challenge to integrating everything into a smoothly working network with service delivery, billing, and management visibility into operations. This remains one of the greatest hurdles, and many see this as a reason why Open RAN and SDN aren't ready for prime time. But the companies that would be called upon to integrate the elements are among the most technically savvy in the world, and their expertise spans telecom network operations and enterprise systems.

No companies are better suited to tackle such a challenge than the large global systems integrators including businesses like Accenture, Capgemini, and Deloitte. Some, like Tech Mahindra, are already investing in the space, and all of them are upskilling their teams to deliver on the requirements for 5G networks, both in traditional configurations and in open, virtualized layouts.

Importantly, this industry sector is dominated by companies based in the United States, Europe, and India, as China has not been able to establish a presence in an industry where success comes through customer intimacy and long-term trust-based agreements.

Jefferson Wang leads Accenture's Global Wireless Practice which includes 5G, Private Networks, Edge Compute, and Advanced Connected Industries. He explains how companies like his can help the US seize the lead in innovation around next-generation mobile solutions.

"The US strength was always collaboration and then differentiation through execution, execution, execution," he says. That collaboration—working across industry sectors or spanning technology platforms—is the bread and butter for the large systems integrators. And Wang sees a way that companies like Accenture can bring a directed purpose to the permission-less innovation environment. "Creativity loves constraints. If you have *too much* choice, sometimes you go in too many directions or you don't have the courage to start."

The systems integrators don't get to dictate how other technology firms spend their time and money, but they can give insight about what is going to be valuable to end customers. Accenture doesn't get paid to create shiny objects—their customers want a functioning worker-safety solution, or a factory-automation package, or a useable autonomous vehicle system. By defining the elements needed to pull this off, companies like Accenture can provide guidance to the many partners in the new ecosystem, letting market opportunities (not central committee dictates) guide investment and development programs.

INTO THE CLOUD

There is another industry sector that is going to have to draw on its know-how and deep pockets to help deliver viable solutions, and it is one in which the US has a huge lead in innovation and scale. The new model for 5G—whether in the Open RAN standard or not—makes a far greater use of the cloud, the remote and centralized concentration of processing power and information. That capability will bring productivity to the abundant, cheap, simple devices that will work as sensors, actuators, and terminals.

From a competitive point of view, this is a very powerful factor. The leading cloud service providers in the US are Amazon, Google, and Microsoft, by far the largest and most successful cloud companies in the world. To the extent that 5G is powered by the cloud, America moves back into a leadership role. The cloud companies are also reportedly exploring partnerships with the tower companies who own the rights-of-way that are needed to deploy the macro networks, creating the possibility of new partnerships and business models to deliver 5G services.

THE NEW PLAYERS

While all these companies are entering the new 5G ecosystem, an industry is developing of smaller companies who are proposing innovative solutions that go beyond what the established equipment makers offer and will be essential to creating the new 5G world that is promised.

ROCKLAND

RED TABLE WINE

ROCKLAND ROAD CELLARS

Cellared And Bottled By

1996

Petite Sirah

NAPA VALLEY

CALISTOGA, CA

750 ML

GOVERNMENT WARNING: (1) ACCORDING TO THE SURGEON GENERAL, WOMEN SHOULD NOT DRINK ALCOHOLIC BEVERAGES DURING PREGNANCY BECAUSE OF THE RISK OF BIRTH DEFECTS. (2) CONSUMPTION OF ALCOHOLIC BEVERAGES IMPAIRS YOUR ABILITY TO DRIVE A CAR OR OPERATE MACHINERY, AND MAY CAUSE HEALTH PROBLEMS.

CONTAINS SULFITES

FOR INFORMATION CALL (707) 963-7439

Some are addressing the need for innovation in radios which, in an Open RAN environment, can be sourced from any equipment maker, swapped for any competing solution, connected to any other vendor's Open RAN boxes, and loaded with any company's software. These flexible solutions can only be supplied if the proprietary handcuffs are broken and more open interfaces are forced on the industry.

Likewise with companies making antennas, which have become a sophisticated value-adding element of the network. 5G's new phased-array antennas create opportunities to enhance service and cut costs, and there are small companies now introducing their own innovative answers.

What do these companies look like? Who is funding them?

More than one of them is being run, or was founded, by an executive who has been around since the industry began, an executive who got his start in wireless some forty years ago, sitting in the back of the conference room and listening to McKinsey declare that the mobile communications market was dead.

Jim Brewington, out of Lucent since 2006 and now in his seventies, is a leader at three companies delivering solutions that he hopes will take the market back from untrusted suppliers. Brew is building teams made up of his old Bell Labs colleagues and twentysomething millennial and Gen Z coders, staffing his companies with people who want to create the next wave of mobile solutions.

"There is not one company that can provide all of 5G," he says. "But even Huawei can't do everything in the 5G space."

Brew founded All Purpose Networks in 2007. The company has an innovative approach to deliver security to IoT networks that have thousands of end points, a large "attack surface." One of the company's products masks the IP address of every device involved in a phone call or data session, from the smartphone or sensor to every router and switch it touches across the network, in effect making communications unhackable, even if they're taking place on a compromised network. Brew says the NSA and the CIA find this intriguing, and the Pentagon sees it as one possible solution in the search for a way to conduct military actions over the networks of allies who may not be as stringent in their rules about whose gear can be deployed into their networks.

Another company he chairs makes innovative base-station radios that can be deployed into Open RAN networks. NewEdge Signal Solutions has been meeting with the big carriers and the government officials who are impressed with their agile products, which use tunable filters and advanced amplifiers to allow more flexibility in how a network is designed and operated. The CEO at NewEdge, Tom Lambalot, says that he's been encouraged to see Democrat and Republican senators cooperating well in their exploration of US-based alternatives to Huawei radios, and the air force has welcomed the technology amid "a dearth of similar companies reaching out."

Brew is also an advisor and investor at Blue Danube, a company that uses beamforming antenna technology, originally developed to shoot down missiles, as a way to make cellular antennas more efficient in delivering bandwidth to smartphones.

Brew has been working to connect his companies to other companies that are well down the road with their own network solutions. One of them, Altiostar, provides the anchor technology for the radical experiment by Japan's Rakuten.

Thierry Maupilé, the EVP of product management and strategy for Massachusetts-based Altiostar, explains the reasoning behind the creation of companies like his. "We saw that 5G would require a different architecture," he says, "more like what was coming from the IT side—the data center—an architecture that could leverage new silicon, software, and cloud. Off-the-shelf hardware at the core and at the edge," instead of the proprietary boxes from the telecom equipment makers. This would not just put the carriers into a better economic path, it has the promise of changing their entire business model, wiping out the costs associated with running the network. "With *automation*, carriers' networks can self-deploy, self-provision, self-configure, and heal themselves when there's a failure." It does far more for the carriers than a cheap, or even a free, network device can do.

But Maupilé sees new architecture as critical for another reason: "The only way you compete against the Chinese is by innovating better and faster. *This* innovation pipeline draws on the US strengths: software, cloud, silicon, artificial intelligence, machine learning. All those technologies are at the heart of what we're doing." In other words, despite the perception that the United States is behind on 5G, by creating a strong and diverse ecosystem, the leaders in our markets can unleash their abilities to become the primary drivers of success.

Another newer company tackling the incumbents is Texas-based Mavenir, which has invested close to $250 million to develop its suite of end-to-end software that controls Open RAN networks and allows the service provider to deliver and manage applications. With more than four thousand employees and over $500 million in revenue, Mavenir is establishing itself as one of the leaders in the category. Chairman Hubert de Pesquidoux explains the advantage he sees his company offering: "When you buy from Huawei, you buy their box, their hardware and software. You have to trust them that it is secure. When you build an Open RAN network, you have open interfaces between all the hardware and software. Outside firms can test everything." Just as important, bugs and vulnerabilities are posted publicly, notifying service providers about what needs to be patched. Stephen Bye, an executive at DISH Network, who is deploying a 5G network in the US that will use Open RAN and SDN elements, put it best in an FCC webinar in 2020: "It's easier to find cockroaches with the lights on." Analyst Roger Entner further points out that the new model allows for automation in the testing of elements for security purposes. This alone adds enormous confidence to the ability to ensure that these complex new networks are safe from compromise. Furthermore, it allows a service provider to very quickly change out any software that is not performing up to standard or presents any other concern.

There is another advantage of this new model for communications. Historically, networks were built to be fail-safe. They had redundant components, back-up elements, extensive pre-deployment testing. Before any device was specified and accepted to go into a network, the service provider had to make certain it did exactly what they said it did; if it didn't, the buyer would be stuck with a faulty device that was integrated into a complete system, a costly problem. With open systems, there will be faulty devices, too, perhaps more of them at the start, but the design is *safe fail*; a problem is not necessarily a catastrophe. When a device doesn't live up to its promise, it can be swapped out for any other vendor's solution, knowing that they all have to work together. Carriers will no longer find the power lies with their vendors, who have the grip of an installed base on their customer.

Today, when a carrier deploys a billion-dollar network from a vendor and finds that one element isn't performing as expected, they are captured. Both buyer and seller know that they can't chuck that one element, which is both integral to the network and proprietary to the vendor.

"The vendors have their hands around the operators' throats," says British-born John Baker, Mavenir's SVP of business development. In the new model, every element can be replaced by anyone offering a competing product. Network components, like network software, will enter a new era of flexibility and agility.

Is this capability really ready to roll? Baker thinks it is. "Four years ago, I would speak about open networks and people would say, 'You're stupid.' Now there's a sense that this could happen. The only piece of custom hardware will be the radio itself, and smaller companies like NewEdge can make that." He also sees this ecosystem reawakening the venture capital community's role in investment, something that has been lacking. "The stagnation of the VC business in telecom gear was because everybody knew that the carriers were only buying from Nokia and Ericsson." With that assumption removed, VCs will be more willing to put their money into network hardware and software makers.

DEVICES AND TERMINALS

This leaves only devices and terminals, already a space where strong competition exists outside of China. While Huawei was able to reach number one status in handsets briefly in 2020, and Xiaomi leads in low-cost devices, the world's leaders in this sector are Samsung and Apple, with Google developing its own devices and supplying critical firmware to make everyone else's devices useful. Their handsets should work fine with the Open RAN interface, and these market-driven companies can be expected to supply whatever the network operators and end users need.

Most of the companies springing up to deliver solutions have management teams based in the United States, whether originally from the country or drawn to a market that has historically provided the best opportunity for entrepreneurs to raise capital, build a team, and reap the financial benefits their success brings. This is a market with the world's broadest supply of technical talent, not just because of the skilled engineers coming out of the country's universities but because of the ability to import or outsource skilled talent from India, Ukraine, Brazil, and, yes, China.

The solution to the 5G challenge lies in turning loose this diverse group of motivated, educated engineers and marketers, and doing so in an environment that has removed the barriers to entry that led to domination by a few large-scale suppliers. Once this new ecosystem is seated on firm ground, service providers, governments, and enterprises across the world will be able to deploy flexible networks that deliver the services people need: factory automation, driverless cars, remote medical care, and other services we haven't yet imagined. Success will be determined by who is more innovative, who provides the most flexibility, freedom, and desired features. Solutions will be dictated not by government authorities but by the desires of the market.

These actions won't occur in a vacuum, of course. China will be fighting every step of the way to retain its leadership in 5G. China is a country where the ruling CCP dictates long-term objectives and pursues them relentlessly: five-year plans, twenty-year projects, hundred-year marathons. This approach has served China well for decades, but left it vulnerable to the agility of free countries and marketplaces, and China may find it has been caught in its own trap. Businesses and industries in free-market societies typically rely on scenario planning, imagining the range of possible outcomes and then shifting direction as circumstances unfold, abandoning plans altogether if a better path emerges.

Authoritarian and totalitarian societies determine the "right answer" and command their subjects to "march forward!" But it appears that China may have instead backed itself into a corner. China made its investment for 5G; it poured tens of billions of dollars into a cheap, effective solution that it plans on rolling out throughout the world over the coming decades. The approach was carefully scoped and executed. The plan was a good one.

The circumstances may have shifted.

If a diverse ecosystem of innovative companies can provide a superior way to deliver flexible, feature-rich services over 5G, China's network equipment makers may find themselves locked into a losing model, compelled to stay on track by their unbending CCP masters as the rest of the world pivots and moves past them. The liberating flexibility of open markets, based on trust and the free exchange of goods in a dynamic ecosystem, will enable free-thinking countries to surge ahead and recapture the leadership of the critical market, returning mobile networks, and all that depends on them, to trusted hands.

EPILOGUE

As I was preparing to send in my manuscript for this book, I received an unusual message over LinkedIn. I was just completing the chapter on how China aggressively recruited and hired critics, ranging from the secretary of the navy to think-tank analysts, when I saw a message from an executive recruiter who wanted to speak with me about a position with a Chinese telecom company. Alarm bells started ringing.

In an earlier conversation with one of my sources for the book, the man who discovered the hack of Nortel's network, he discussed the risks of writing this and keeping it safe from the prying eyes of nation-state hackers. "Do the whole book on a standalone, nonnetworked computer," he warned. "Save it to memory sticks. Keep them in a safe-deposit box." He related how he was giving a public presentation about his hacking discoveries when a Chinese dissident approached him and asked to get in touch. He set up a Yahoo email account for the exchange, securing it with an unguessable password. They traded emails, and the next week his account was accessed and his password reset. He learned that Yahoo's email system was compromised by China a year before Yahoo knew.

I decided that I couldn't put a year's worth of work on a stick and a hard drive and risk losing everything. I also figured that a state actor, if they were interested, would get past any security I attempted. I assumed that, if they cared, whatever I was writing, they would be reading.

But this LinkedIn message came out of the blue. I responded to the recruiter and asked to see the job description. The position they were looking to fill was for a board seat on China Unicom Americas, a division of the massive state-owned telecom operator in China. They have provided internet connectivity to North Korea since 2010, though I don't believe they use that fact in their marketing campaigns. The job's main responsibilities would be to chair the board's risk management committee and advise them on how to respond to a Department of Justice inquiry on the risk they posed to US national security. It was a new position, and I would be the only non-Chinese, non-Unicom executive on the board.

I asked how they found me and was told it was through a standard LinkedIn search, where I showed up as an alum of the same school as the recruiter, with a background in telecom. I had no mention of China on my LinkedIn page, no experience representing a company in front of the Department of Justice, let alone a subsidiary of a company with more customers than AT&T and Verizon combined.

It didn't make sense.

The recruiter called me, and we spoke about the job. I offered that I knew someone who was much better suited to this position. He had just left a job running government affairs in the United States for a major Chinese technology company, knew telecom inside and out, and was well respected by the US government officials he would be dealing with. He had counseled his Chinese employer on exactly these issues.

She didn't want his name.

Before we ended our call, I had one last question. This was the first position I had ever seen that didn't provide a salary. It didn't even mention a range. I asked what they planned to pay.

She said it was negotiable.

A NOTE REGARDING SOURCES

The stories, assertions, and recollections in this book come from nearly one hundred interviews with executives; scientists; academics; cybersecurity experts; and intelligence, law enforcement, and other government officials. They also come from my direct experiences working in the industry; more than once, I approached an interview subject with my own recollections and asked them to corroborate my memory or set the record straight. The events described also draw on information in the public record, as detailed in the endnotes. Any quotations not endnoted are drawn from my direct conversations with the subjects of the story. Every scene described has been corroborated by multiple people or is supported by contemporaneous evidence or news accounts.

The dialog presented is reconstructed from the memories of the subjects and validated by other parties who were present whenever possible. The names used are the subjects' real names in almost every case; in one case a minor figure's name is lost to history and in four other anecdotes the subjects requested that their real names not be used due to their role in the intelligence community or due to sensitivities around commercial contracts. In each of those cases the subject was known to the author, and his legitimacy as a source of valid information was independently confirmed.

ACKNOWLEDGMENTS

There are two people, above all, that I want to thank for their role in this book. One is my brother Mark, who proposed the idea, and kept at me for some time until I decided to write it. The other is my wife, Katie. When I floated the idea, she could have told me it was foolish; instead, she encouraged me to pursue this, knowing that I had always wanted to write. Katie, I would have known if you were just saying it to make me happy. Thank you.

The encouragement and input from other family and friends were critical, from my kids, Mary and Ben, who had constant advice and assistance, to my brother Dan and my mother, whose support was unwavering.

To my friends, and talented writers, Eric Dean and Richard Longstaff, thank you for the time and effort you put into reviewing my earliest drafts.

The other people who made the idea a reality were Daniel Greenberg, my agent at Levine Greenberg Rostan, who didn't hesitate to sign me despite being a first-time author, and Glenn Yeffeth, my publisher at BenBella, who was flexible and innovative in putting a deal together. You gave me the chance. Josh Dean, who introduced me to his agent, spent time he didn't have, and helped me prepare a proposal, using his own capital to put me in front of the best agent in the industry. Vy Tran, my editor, worked tirelessly to make this flow, focus on the important parts, and bring me out of the insider lingo; Scott Calamar, my copy editor, was meticulous, encouraging, and clever with his work on the book.

There were over one hundred interviews that went into the book, and for everyone whose name appears within, thank you for sharing stories and insights that people needed to know. Jim Brewington, above all, gave more time and attention than I could have hoped, and was a reality check on much of my writing. Dan Hesse, too, gave time and made introductions that were critical in helping me reach other leaders in the business. Colin Golder and Ron Merino also gave enormous time, and their stories were essential for the book.

For the others who gave me technical or commercial information on the subject matter, thank you for your contributions. Thanks especially to Eric Burger and Roger Entner for reviewing technical sections and fixing things I got wrong. (I kept writing after your reviews, so if any technical mistakes made it into the book, that's on me.) For the people in the intelligence community and law enforcement, and in other branches of government, thanks for the inside scoop. You know who you are.

Cliff May at FDD was kind enough to connect me with some of the best experts in the world on policy, security, and strategy.

Many other people sat for interviews and, although their names don't appear herein, their insight and guidance were critical in helping me understand the story. Some of your companies are skittish about seeing their executives quoted in a book about China, so I will spare you the grief of thanking you by name, but I am grateful.

This was a story that needed to be told, and all of you helped me tell it.

NOTES

PROLOGUE

xiii   **These systems were:** Nancy Gan, "With the Coronavirus Under Control, This Chinese City Wants to Score and Rank Its Residents Based on Their Health and Lifestyle," *CNN Business*, May 25, 2020, https://www.cnn.com/2020/05/25/tech/hangzhou-health-app-intl -hnk/index.html

xiii   **Cash payment was:** Andrey Shevchenko, "China Tracks Victims of Coronavirus with WeChat and Alipay," *Cointelegraph*, February 18, 2020, https://cointelegraph.com/news /china-tracks-victims-of-coronavirus-with-wechat-and-alipay

xiii   **The information was:** Helen Davidson, "China's Coronavirus Health Code Apps Raise Concerns over Privacy," *Guardian*, April 1, 2020, https://www.theguardian.com /world/2020/apr/01/chinas-coronavirus-health-code-apps-raise-concerns-over-privacy

xiv   **Even Wuhan:** Hannah Ritchie, et al., "Mortality Risk of COVID-19," *Our World In Data*, accessed February 8, 2021, https://ourworldindata.org/mortality-risk-covid#the-case -fatality-rate; Kai Kupferschmidt and Jon Cohen, "China's Aggressive Measures Have Slowed the Coronavirus. They May Not Work in Other Countries," *Science Magazine*, March 2, 2020, https://www.sciencemag.org/news/2020/03/china-s-aggressive-measures -have-slowed-coronavirus-they-may-not-work-other-countries

xiv   **It appeared that:** David Brunnstrom, "U.S. Appeals to China to Revise Export Rules on Coronavirus Medical Gear," Reuters, April 16, 2020, https://www.reuters.com/article /us-heath-coronavirus-usa-china/u-s-appeals-to-china-to-revise-export-rules-on -coronavirus-medical-gear-idUSKBN21Z07G

xiv   **Yet on January 30:** "WHO Director-General's Statement on IHR Emergency Committee on Novel Coronavirus (2019-nCoV)," World Health Organization, January 30, 2020, https://www.who.int/director-general/speeches/detail/who-director-general-s-statement -on-ihr-emergency-committee-on-novel-coronavirus-(2019-ncov)

xiv   **The Associated Press:** "China Delayed Releasing Coronavirus Info, Frustrating WHO," Associated Press, June 2, 2020, https://apnews.com/article/3c061794970661042b18d5ae aaed9fae

xiv   **But when Australia's:** "Australia Accuses China of Undermining Trade Agreement," *BBC News*, December 9, 2020, https://www.bbc.com/news/world-australia-55240898

xv   **Concerns had been raised:** Rich Haridy, "Huawei, the US Ban, and Links to Chinese
       Spying Explained," *New Atlas*, May 22, 2019, https://newatlas.com/huawei-ban-us-what
       -spy-evidence-exists/59772/

xv   **As some countries:** Shi Jiangtao, "Chinese Ambassador Accused of Threatening German
       Car Industry if Huawei Is Frozen Out," *South China Morning Post*, December 15, 2019,
       https://www.scmp.com/news/china/diplomacy/article/3042190/chinese-ambassador
       -accused-threatening-german-car-industry-if

INTRODUCTION

2   **But Huawei has grown:** Nathaniel Ahrens, "China's Competitiveness Myth, Reality, and
       Lessons for the United States and Japan," Center for Strategic & International Studies,
       February 2013, https://csis-website-prod.s3.amazonaws.com/s3fs-public/legacy_files/files
       /publication/130215_competitiveness_Huawei_casestudy_Web.pdf

2   **Bigger than all the other:** "Telecommunications Equipment Companies Ranked by
       Overall Revenue in 2018," Statista, https://www.statista.com/statistics/314657/top-10
       -telecom-equipment-companies-revenue/

2   **While also surpassing:** Zak Dorffman, "Samsung and Apple Beaten by Huawei in Huge
       New Smartphone Surprise," *Forbes*, June 15, 2020, https://www.forbes.com/sites/zak
       doffman/2020/06/15/samsung-and-apple-beaten-by-huawei-in-huge-new-smartphone
       -surprise/?sh=2b8fa98a58ab

2   **Their twenty billion:** Dan Strumpf, "Huawei Workers Return After Coronavirus, But
       CEO Sees Financial Hit," *Wall Street Journal*, March 25, 2020, https://www.wsj.com/articles
       /huawei-workers-return-after-coronavirus-but-ceo-sees-financial-hit-11585149231

3   **Air Force global:** FCC 19-121 Statement from Commissioner Jessica Rosenworcel, "Re: Pro-
       tecting Against National Security Threats to the Communications Supply China Through
       FCC Program, WC Docket No 18-89," November 26, 2019, https://docs.fcc.gov/public
       /attachments/FCC-19-121A5.pdf

3   **Far too many:** Alexander Martin, "Huawei: The Company and the Security Risks
       Explained" *Sky News*, September 23, 2020, https://news.sky.com/story/huawei-the
       -company-and-the-security-risks-explained-11620232

3   **Countries around the world:** Justin Sherman, "Is the U.S. Winning Its Campaign
       Against Huawei?" *Lawfare* (blog), August 12, 2020, https://www.lawfareblog.com/us
       -winning-its-campaign-against-huawei

4   **German chancellor Angela:** Ben Knight, "Angela Merkel Faces Party Revolt over Huawei
       in German 5G Rollout," DW.com, November 22, 2019, https://www.dw.com/en/angela
       -merkel-faces-party-revolt-over-huawei-in-german-5g-rollout/a-51372875

4   **Other national leaders:** Kate Proctor, "Huawei: Government Tries to Head Off Tory
       5G Network Rebellion," *Guardian*, March 9, 2020, https://www.theguardian.com
       /technology/2020/mar/09/huawei-government-tries-to-head-off-5g-network-rebellion

4   **And the world's:** Joe Gould, "'Nancy Pelosi and Donald Trump See Huawei the Same.' 5G
       in Europe aligns America's top political rivals," Yahoo!finance, February 14, 2020, https://
       finance.yahoo.com/news/nancy-pelosi-donald-trump-see-184500716.html

6   **Take Lucent:** Tom von Alten, "Top 100 Companies by Market Capitalization," Fort Boise,
       December 28, 2001, https://fortboise.org/top100mktcap.html

6   **together valued:** World Top 1000 Companies List and World Ranks, accessed January 1,
       2021, https://www.value.today/

7   **Credible reports:** Matthew Hill, David Campanale, and Joel Gunter, "'Their Goal Is to Destroy Everyone': Uighur Camp Detainees Allege Systematic Rape," *BBC News*, February 2, 2021, https://www.bbc.com/news/world-asia-china-55794071

7   **For example, Chinese:** Lucy Craymer, "China's National-Security Law Reaches into Harvard, Princeton Classrooms," *Wall Street Journal*, August 19, 2020, https://www.wsj.com/articles/u-s-colleges-are-taking-steps-to-protect-students-from-chinas-national-security-law-11597954115

8   **Or does it:** "Our Culture," Huawei corporate web site, April 19, 2021, https://huaweico.wordpress.com/our-culture/

8   **As Huawei deploys:** Katherine Atha, et al., "China's Smart Cities Development," U.S.-China Economic and Security Review Commission, January 2020, https://www.uscc.gov/sites/default/files/2020-04/China_Smart_Cities_Development.pdf

8   **watchdog groups:** Yau Tsz Yan, "Smart Cities or Surveillance? Huawei in Central Asia," *The Diplomat*, August 7, 2019, https://thediplomat.com/2019/08/smart-cities-or-surveillance-huawei-in-central-asia/

8   **Activists and officials:** Paul Mozur, Jonah M. Kessel, and Melissa Chan, "Made in China, Exported to the World: The Surveillance State," *New York Times*, April 24, 2019, https://www.nytimes.com/2019/04/24/technology/ecuador-surveillance-cameras-police-government.html/

8   **Strong-arm governments:** Bulelani Jili, "The Spread of Surveillance Technology in Africa Stirs Security Concerns," Africa Center for Strategic Studies, December 11, 2020, https://africacenter.org/spotlight/surveillance-technology-in-africa-security-concerns/

8   **Huawei denies the claims:** Daniele Lepido, "Vodafone Found Hidden Backdoors in Huawei Equipment," *Bloomberg*, April 30, 2019, https://www.bloomberg.com/news/articles/2019-04-30/vodafone-found-hidden-backdoors-in-huawei-equipment

8   **Huawei asserts their:** Matt Hamblen, "Huawei Executive Claims Its Private-Company Status Means It's Not Controlled by China's Government," *Fierce Wireless*, June 5, 2019, https://www.fiercewireless.com/wireless/huawei-exec-claims-its-private-company-status-means-it-s-not-controlled-by-china-s

8   **But recent research:** Murray Scot Tanner, "Beijing's New National Intelligence Law: From Defense to Offense," *Lawfare* (blog), July 20, 2017, https://www.lawfareblog.com/beijings-new-national-intelligence-law-defense-offense

9   **The United States:** Stu Woo, "Facing Pushback from Allies, US Set for Broader Huawei Effort," *Wall Street Journal*, January 23, 2020, https://www.wsj.com/articles/facing-pushback-from-allies-u-s-set-for-broader-huawei-effort-11579775403

9   **China recognized this:** Kathrin Hille, Nic Fildes, and Qianer Liu, "US-China: Is Huawei 'Too Big to Fail'?" *Financial Times*, August 21, 2020, https://www.ft.com/content/2c378685-e04d-40cc-b986-7eef594c7255

CHAPTER 1: THE STUDY

15   **Market potential for cellular:** Presentation from McKinsey & Co. to AT&T Technologies, c. 1986

15   **Although some stories:** Archivist, The Centre for Business History in Stockholm, collected January 2021, https://www.naringslivshistoria.se/en/

16 **As sales slowed:** Arnold Rombo, "Innovation strategies of the 19th Century, The case of LM Ericsson," Masters Thesis, Linkoping University, 2011, https://www.diva-portal.org/smash/get/diva2:446632/FULLTEXT03

16 **hiring aggressive overseas:** "Axel Bostrom, 1900-1909," Ericsson company archives, https://www.ericsson.com/en/about-us/history/people/presidents/axel-bostrom-1900-1909

16 **By covering the:** "The Foundations of Mobile and Cellular Telephony," From Today's Engineer, as reported in Engineering and Technology History Wiki, August 2012, https://ethw.org/The_Foundations_of_Mobile_and_Cellular_Telephony

17 **AT&T was a:** Tim Wu, "A Brief History of American Telecommunications Regulation," Columbia Law School Scholarship Archive, 2007, https://scholarship.law.columbia.edu/cgi/viewcontent.cgi?article=2462&context=faculty_scholarship

17 **On April 3:** Marty Cooper, interview by Richard Taylor, "Interview with Mobile Phone Inventor Marty Cooper," *BBC Click*, https://www.bbc.com/news/av/technology-22020666

18 **That turned out:** The Study actually consisted of a number of projects and presentations from 1980 through the mid-eighties. The deck cited is undated but was part of this series and would have been presented in 1985 or 1986.

21 **While AT&T was:** "Cell Phone Development," Motorola Solutions, https://www.motorolasolutions.com/en_us/about/company-overview/history/explore-motorola-heritage/cell-phone-development.html

21 **He pronounced his:** Heinrich Hertz, "Electric Waves," 1893, https://archive.org/details/electricwavesbe00hertgoog/page/n4/mode/2up

21 **By 2000 there:** "Number of Mobile (Cellular) Subscriptions Worldwide from 1993 to 2019," Statista, https://www.statista.com/statistics/262950/global-mobile-subscriptions-since-1993/

21 **By 2018 more:** Damian Radcliffe, "Mobile in Sub-Saharan Africa: Can World's Fastest-Growing Mobile Region Keep It Up?" *ZDNet*, October 16, 2018, https://www.zdnet.com/article/mobile-in-sub-saharan-africa-can-worlds-fastest-growing-mobile-region-keep-it-up/

22 **Penetration by 2020:** S. O'Dea, "Number of Mobile Subscriptions Worldwide 1993-2019," Statista, December 3, 2020, https://www.statista.com/statistics/262950/global-mobile-subscriptions-since-1993/

23 **AT&T missed an:** Andrew Kupfer, "AT&T's $12 Billion Cellular Dream," CNN Money, December 12, 1994, https://money.cnn.com/magazines/fortune/fortune_archive/1994/12/12/80051/index.htm

23 **Instead, Brew suffered:** Michael Kanellos, "AT&T Has Troubled Past in Mergers," CNet, June 24, 1998, https://www.cnet.com/news/at-t-has-troubled-past-in-mergers/

23 **If AT&T:** Nokia Bell Labs, accessed April 20, 2021, https://www.bell-labs.com/timeline/#/2010/1/closed/

23 **Their scientists had:** Nokia Bell Labs, accessed April 20, 2021, https://www.bell-labs.com/about/awards/2018-nobel-prize-physics/

24 **Large sections of:** Jonathan Waldman, *Rust, The Longest War* (New York: Simon & Schuster, 2016), https://mmsallaboutmetallurgy.com/wp-content/uploads/2018/12/War-against-Rust.pdf

24 **With over 827:** John S. DeMott and Bruce van Voorst, "Click! Ma Is Ringing Off," *Time* magazine, November 21, 1983, https://www.beatriceco.com/bti/porticus/bell/pdf/1983_11_21_time_splitting_att.pdf

24   **Franey spent the:** Rick Hampson, "Old Copper Roof Gives Statue of Liberty a New Dress," AP, October 11, 1985, https://apnews.com/article/b997ef41c27c3e0fa5690b0bcfa7bdb6

26   **Thirty years later:** Douglas Natelson and Don Monroe, "End of an Era," *nanoscale views* (blog), April 13, 2014 http://nanoscale.blogspot.com/2014/04/end-of-era.html

26   **A relatively successful:** Matthew Christopher, "Bell Labs," *Abandoned America* (blog), June 23, 2019, https://www.abandonedamerica.us/bell-labs

CHAPTER 2: AT&T CREATES A BUSINESS

27   **Always remember that:** "It's Easier to Ask Forgiveness Than to Get Permission," Quote Investigator, https://quoteinvestigator.com/2018/06/19/forgive/

29   **There, Bell Labs:** Michael Noll, "Memories: A Personal History of Bell Telephone Laboratories," Michigan State University quello center working paper, August 6, 2015, https://quello.msu.edu/wp-content/uploads/2015/08/Memories-Noll.pdf

CHAPTER 3: RETAKING THE LEAD

31   **Creating the future:** Gary Hamel, "Competing for the Future—Breakthrough Strategies for Seizing Control of Your Industry and Creating Markets of Tomorrow," *Harvard Business Review*, July-August 1994, https://hbr.org/1994/07/competing-for-the-future

33   **By 2002, when:** "Italy Teledensity: Mobile," CEIC Data, accessed April 21, 2021, https://www.ceicdata.com/en/indicator/italy/teledensity-mobile

CHAPTER 4: LOSING THE PLOT

37   **You can have:** "Steffi Graf Quotes," BrainyQuote, https://www.brainyquote.com/quotes/steffi_graf_480401

37   **A graduate of:** Leslie Cauley, "CEO Profile: Sprint's New CEO Showed Grit from the Start," *USA Today*, March 3, 2008, https://abcnews.go.com/Business/CEOProfiles/story?id=4338159&page=1

40   **In December of:** Reily Gregson, "Intercepted Gingrich Cell Phone Call May Lead to Civil Lawsuit," *RCRWireless*, November 17, 1997, https://www.rcrwireless.com/19971117/archived-articles/intercepted-gingrich-cell-phone-call-may-lead-to-civil-lawsuit

CHAPTER 5: POVERTY AND THE POWER OF COMMUNICATION

42   **It doesn't matter:** "It doesn't matter if a cat is black or white, as long as it catches mice," *The Oxford Dictionary of Proverbs* (5 ed.), edited by John Simpson and Jennifer Speake (Oxford, UK: Oxford University Press, 2009), https://www.oxfordreference.com/view/10.1093/acref/9780199539536.001.0001/acref-9780199539536-e-312

44   **The Minzu had:** Jack Aldane, "Beijing's Original Skyscraper Short in Height But Not in History," *Global Times*, October 21, 2012, https://www.globaltimes.cn/content/739551.shtml

46   **In 1980, China:** "Here's How Much Poverty Has Declined in China," *Wall Street Journal*, April 17, 2013, https://www.wsj.com/articles/BL-CJB-17580

46   **The problem was:** "China GDP Per Capita 1960–2021," macrotrends, https://www.macro trends.net/countries/CHN/china/gdp-per-capita

46   **Increasing teledensity caused:** Emmanuel Forestier, Jeremy Grace, and Charles Kenny, "Can Information and Communications Policy Be Pro-Poor?" *Telecommunications Policy* 26 (2002): 623–646, https://www.researchgate.net/figure/Quality-of-life-and-teledensity -regression-results_tbl5_227350683

48   **But the country:** Eric Harwit, "Spreading Telecommunications to Developing Areas in China: Telephones, the Internet and the Digital Divide," *The China Quarterly* 180 (December 2004): 1010–1030, https://www.jstor.org/stable/20192415?seq=1

CHAPTER 6: THE PRICE OF ADMISSION

49   **The contract is:** Deb Weidenhamer, "The Contract Is Signed. and Now the Negotiation Begins," *You're the Boss: The Art of Running a Small Business, New York Times* blog, January 28, 2014, https://web.archive.org/web/20191005024831/https://boss.blogs.nytimes.com /2014/01/28/the-contract-is-signed-and-now-the-negotiation-begins/

49   **It wasn't until:** Ernestina Coast, "Population Trends in Developing Countries," LSE Research Online, 2002, http://eprints.lse.ac.uk/268/1/Arnold.pdf

49   **More than 4:** Mobile Cellular Subscriptions, International Telecommunication Union Indicators Database, https://data.worldbank.org/indicator/IT.CEL.SETS

50   **Including the traditional:** Eric Harwit, "Spreading Telecommunications to Developing Areas in China; Telephones, the Internet and the Digital Divide," *The China Quarterly* 180 (December 2004): 1010–1030 https://www.jstor.org/stable/20192415?seq=1

50   **Among the top:** Kathryn Kranhold, "China's Price for Market Entry: Give Us Your Technology, Too," *Wall Street Journal*, February 26, 2004, https://www.wsj.com/articles /SB107775213437639391

50   **Cellular communications was:** World Telecommunication Development Report 199, Mobile Cellular, October 1999, https://www.itu.int/ITU-D/ict/publications/wtdr_99 /material/wtdr99s.pdf

50   **and explosive growth:** Mobile Cellular Subscriptions, International Telecommunication Union, Indicators Database, https://data.worldbank.org/indicator/IT.CEL.SETS

53   **His instincts were:** "Commission on the Theft of American Intellectual Property," National Bureau of Asian Research, February 27, 2017, https://www.nbr.org/program /commission-on-the-theft-of-intellectual-property/

CHAPTER 7: THE BIRTH OF HUAWEI

59   **Since the Opium:** Norman Pearlstine et al, "Who's Behind Huawei?" *Los Angeles Times*, April 10, 2019, https://enewspaper.latimes.com/infinity/article_share.aspx?guid=4e28625b -9cb4-47b3-b21e-1db258a812fc

59   **In 1987, he:** Karishma Vaswani, "Huawei, the Story of a Controversial Company," *BBC News*, March 6, 2019, https://www.bbc.co.uk/news/resources/idt-sh/Huawei

65   **Short of finding:** Christopher Balding and Donald Clarke, "Who Owns Huawei?" *SSRN*, April 17, 2019, https://papers.ssrn.com/sol3/papers.cfm?abstract_id=3372669

66   **There are many:** Elizabeth Redden, "Not Feeling Safe in China," *Inside Higher Ed*, July 23, 2018, https://www.insidehighered.com/news/2018/07/23/american-academic-leaves-china -citing-concerns-about-physical-safety

66   **At one point:** Peter Hogfeldt, "The History and Politics of Corporate Ownership in Sweden," National Bureau of Economic Research, July 2004, https://www.nber.org/papers/w10641

68   **They describe the:** Scott Bicheno, "Huawei Responds to Study Questioning Who Really Owns It," telecoms.com, April 18, 2019, https://telecoms.com/496982/huawei-responds-to-study-questioning-who-really-owns-it/

CHAPTER 8: RISING STAR

71   **These ship-to-ship:** James Martin Center, "Iraq Missile Chronology," Nonproliferation Studies at the Monterey Institute of International Studies, November, 2008, https://media.nti.org/pdfs/iraq_missile.pdf

72   **On his meeting:** "Crown Prince Abdullah Completes Historic World Tour," Monthly Newsletter of the Royal Embassy of Saudi Arabia, Washington, D.C., November 1998, https://books.google.com/books?id=R7EtAQAAIAAJ&pg=RA2-PA2&lpg=RA2-PA2&dq=%22Crown+Prince+Abdullah%22+visit+china+1998+rongji+joint+memorandum&source=bl&ots=0DxqyorC6K&sig=ACfU3U05pTmD3_Muo1SsM1sCP_WUUIsfMw&hl=en&sa=X&ved=2ahUKEwjqut_HgOruAhWKF1kFHUjyDLcQ6AEwBH0ECAgQAg#v=onepage&q=%22Crown%20Prince%20Abdullah%22%20visit%20china%201998%20rongji%20joint%20memorandum&f=false

75   **That might have:** Alumni Newsletter, University of Electronic Science and Technology of China, April 3, 2006, http://web.archive.org/web/20070607165123/http://www.alumni.uestc.edu.cn/ListNews.aspx?ID=484

CHAPTER 9: BOOM TO BUST

78   **Never Interrupt your:** "Never Interfere with an Enemy While He's in the Process of Destroying Himself," Quote Investigator, https://quoteinvestigator.com/2010/07/06/never-interfere/

79   **Their stocks soared:** Jonathan Calof et al, "An Overview of the Demise of Nortel Networks and Key Lessons Learned," Telfer School of Management," 2014, https://sites.telfer.uottawa.ca/nortelstudy/files/2014/02/nortel-summary-report-and-executive-summary.pdf

79   **Nokia's stock comprised:** Gordon Kelly, "Finland and Nokia: An Affair to Remember," *Wired*, April 10, 2013, https://www.wired.co.uk/article/finland-and-nokia

81   **In 2001, Lucent:** Lucent Technologies Annual Report, 2002, https://beatriceco.com/bti/porticus/bell/pdf/lucent_2002.pdf

81   **financed nearly $8 billion:** "Vendor Financing," *LightReading*, December 6, 2000, https://www.lightreading.com/ethernet-ip/vendor-financing/d/d-id/571917

81   **Lucent would build:** "Lucent Sunk Cash in Golf Club But Wants to Play Thru," NewJersey Hills.com, February 1, 2001, https://www.newjerseyhills.com/lucent-sunk-cash-in-golf-club-but-wants-to-play-thru/article_1abade83-06b6-5f05-93be-f9e617481a5a.html

82   **The market cap:** William Lazonick and Edward March, "The Rise and Demise of Lucent Technologies," The AIR Net, March 2010, http://www.theairnet.org/files/research/lazonick/Lazonick%20and%20March%20Lucent%20COMPLETE%2020110324.pdf

82   **The company's facilities:** Carol J. Loomis, "The Whistleblower and the CEO," *CNN Money*, July 7, 2003, https://money.cnn.com/magazines/fortune/fortune_archive/2003/07/07/345538/index.htm

83   **By 2001, as Huawei:** Huawei Technologies Statistics and Facts, Market.us, https://market.us/statistics/smartphone-brands/huawei/

83   **By 2001, Lucent:** Lucent Technologies Annual Report, 2002, https://beatriceco.com/bti/porticus/bell/pdf/lucent_2002.pdf

83   **The telecom equipment:** Paul Starr, "The Great Telecom Implosion," *American Prospect*, September 8, 2002, https://www.princeton.edu/~starr/articles/articles02/Starr-Telecom Implosion-9-02.htm

84   **Capital Spending by:** Dennis K. Berman, "Worst 3, 5-Year Performer: Lucent," *Wall Street Journal*, March 10, 2003, https://www.wsj.com/articles/SB104699082198862800

84   **In their filings:** Reily Gregson, "Winstar Bankruptcy, Lawsuit Detail Ugly Side of Vendor Financing," *RCRWireless*, April 23, 2001, https://www.rcrwireless.com/20010423/carriers/winstar-bankruptcy-lawsuit-detail-ugly-side-of-vendor-financing

CHAPTER 10: DEAL OR NO DEAL

87   **He roughs up:** Adam Lashinsky, "Shut Up and Deal," *Wired*, March 3, 2001, https://www.wired.com/2001/03/boutros/

89   **Alcatel was announcing:** John Tagliabue, "Shares of Alcatel Fall on Profit Warning," *New York Times*, May 31, 2001, https://www.nytimes.com/2001/05/31/business/shares-of-alcatel-fall-on-profit-warning.html

CHAPTER 11: THE SINCEREST FORM OF FLATTERY

92   **According to the report:** Lucent Bell Labs Competitive Analysis, https://wenku.baidu.com/view/366283f6f61fb7360b4c65ac?pcf=2

93   **Several criminal and:** "Chinese Telecommunications Conglomerate Huawei and Subsidiaries Charged in Racketeering Conspiracy and Conspiracy to Steal Trade Secrets," US Department of Justice, February 13, 2020, https://www.justice.gov/opa/pr/chinese-telecommunications-conglomerate-huawei-and-subsidiaries-charged-racketeering; Jim Duffy, "Cisco Sues Huawei over Intellectual Property," Computerworld, January 23, 2003, https://www.computerworld.com/article/2578617/cisco-sues-huawei-over-intellectual-property.html

94   **The court filings:** "Declaration of Chad Reynolds Supplementing Cisco's motion for preliminary injunction," March 14, 2003, https://newsroom.cisco.com/dlls/declaration_chad_reynolds.pdf

94   **Huawei settled with:** Mark Chandler, "Huawei and Cisco's Source Code: Correcting the Record," *Cisco Blogs,* October 11, 2012, https://blogs.cisco.com/news/huawei-and-ciscos-source-code-correcting-the-record

95   **After obtaining a:** "New Indictment Expands Charges Against Former Lucent Scientists Accused of Passing Trade Secrets to Chinese Company," US Department of Justice

Press Release, April 11, 2002, https://www.justice.gov/archive/criminal/cybercrime/press-releases/2002/lucentSupIndict.htm

96   **There was, however:** Jeff Baumgartner, "Lucent: PathStar Gone, VoIP Lives," NextTV, February 19, 2001, https://www.nexttv.com/news/lucent-pathstar-gone-voip-lives-134286

CHAPTER 12: A FOOTHOLD IN EUROPE

99   **When he graduated:** "Ben Verwaayen," Reference for Business, https://www.referenceforbusiness.com/biography/S-Z/Verwaayen-Ben-1952.html

99   **His next job:** "Papers Show I.T.T. Urged US to Help Oust Allende," *New York Times*, July 3, 1972, https://www.nytimes.com/1972/07/03/archives/papers-show-itt-urged-us-to-help-oust-allende-suggestions-for.html

100   **The hundred-year-old:** "Marconi Plunges on BT Deal News," *BBC News*, April 28, 2005, http://news.bbc.co.uk/2/hi/business/4492581.stm

100   **By October, 2005:** Saeed Shah, "End of an Era as US Firm Buys Marconi Rump," *Independent*, October 23, 2011, https://www.independent.co.uk/news/business/news/end-of-era-as-us-firm-buys-marconi-rump-479791.html

101   **In 2010:** Andrew Probyn, "Huawei's History in Britain May Help Explain Why Australia Is So Nervous," Australian Broadcasting Corporation, June 16, 2018, https://www.abc.net.au/news/2018-06-16/huawei-britain-history-helps-explain-australia-anxiety/9875582

102   **The committee sought:** "Foreign Involvement in the Critical National Infrastructure: The Implications for National Security," Intelligence and Security Committee, June 2013, https://assets.publishing.service.gov.uk/government/uploads/system/uploads/attachment_data/file/205680/ISC-Report-Foreign-Investment-in-the-Critical-National-Infrastructure.pdf

103   **Not long after:** Amit Katwala, "Here's How GCHQ Scours Huawei Hardware for Malicious Code," *Wired*, February 22, 2019, https://www.wired.co.uk/article/huawei-gchq-security-evaluation-uk

104   **In the 2020:** "Huawei Cyber Security Evaluation Centre (HCSEC) Oversight Board Annual Report 2019," March 2019, https://assets.publishing.service.gov.uk/government/uploads/system/uploads/attachment_data/file/790270/HCSEC_OversightBoardReport-2019.pdf

CHAPTER 13: SIZING UP HUAWEI AND
THE GLOBAL MARKETPLACE

107   **It is a myth:** "Sony Ericsson Makes China Its Global Manufacturing Base," *China Daily*, June 19, 2007, https://www.chinadaily.com.cn/bizchina/2007-06/19/content_897613.htm

109   **Yet according to:** "We See Beyond Telecom," Huawei Technologies Co., Ltd., 2010 Annual Report, Huawei, https://www.huawei.com/ucmf/groups/public/documents/annual_report/092576.pdf

109   **Kim went on:** Iain Morris, "Samsung: Huawei Pricing Not Viable for Profit-Seeking Company," *LightReading*, July 9, 2020, https://www.lightreading.com/5g/samsung-huawei-pricing-not-viable-for-profit-seeking-company/d/d-id/762293

109   **Reports from the:** Dan Steinbock, "The Case for Huawei in America," September 3, 2012, https://huawei.mediaroom.com/download/20120904+Case+for+Huawei+in+America-Huawei+%28ds%29.pdf

110 **By the time:** Saul Hansell, "Technology: A Surprise from Amazon: Its First Profit," *New York Times*, January 23, 2002, https://www.nytimes.com/2002/01/23/business/technology-a-surprise-from-amazon-its-first-profit.html

110 **In a report:** Chuin-Wei Yap and Matthew Dalton, "State Support Helped Fuel Huawei's Global Rise," *Wall Street Journal*, December 25, 2019, https://www.wsj.com/articles/state-support-helped-fuel-huaweis-global-rise-11577280736

110 **Founder and CEO:** "Ren Zhengfei's Interview with BBC," Huawei, February 18, 2019, https://www.huawei.com/us/facts/voices-of-huawei/ren-zhengfei-interview-with-bbc

110 **the company received:** Ryan Mcmorrow, "Huawei a Key Beneficiary of China Subsidies That US Wants Ended," Phys.org, May 30, 2019, https://phys.org/news/2019-05-huawei-key-beneficiary-china-subsidies.html

111 **A December, 2000:** Bruce Gilley, "Huawei's Fixed Line to Beijing," *Far Eastern Economic Review*, http://www.web.pdx.edu/~gilleyb/Huawei_FEER28Dec2000.pdf

CHAPTER 14: THE END—FOR NOW

115 **For China, purely:** Attorney General William Barr, keynote address at the Department of Justice's China Initiative Conference, https://www.justice.gov/opa/speech/attorney-general-william-p-barr-delivers-keynote-address-department-justices-china

116 **The company entered:** John Shinal, "Alcatel to Buy Lucent for $13.5 billion," *Market Watch*, April 3, 2006, https://www.marketwatch.com/story/alcatel-to-acquire-lucent-for-135-billion-in-stock

116 **The offer was:** Carole J Loomis, "The Whistleblower and the CEO," *Fortune*, July 7, 2003, http://money.cnn.com/magazines/fortune/fortune_archive/2003/07/07/345538/

117 **Their role is:** U.S. Department of the Treasury, The Committee on Foreign Investment in the United States (CFIUS), https://home.treasury.gov/policy-issues/international/the-committee-on-foreign-investment-in-the-united-states-cfius

CHAPTER 15: CULTURE AND CREDIT CARDS

123 **He who does:** "Lao Tzu Quotes," BrainyQuote, https://www.brainyquote.com/quotes/lao_tzu_379183

127 **Revenue for 2008:** "Enriching Life Through Communication," Annual Report 2008, Huawei, https://www.huawei.com/ucmf/groups/public/documents/annual_report/092581.pdf

CHAPTER 16: BRAIN DRAIN

130 **Trust but verify:** Byron Rashed, "Trust but verify . . ." *Centripetal Blog*, July 28, 2020, https://www.centripetal.ai/blog/trust-but-verify

131 **The company hiring them:** Futurewei, accessed April 20, 2021, https://www.futurewei.com/

133 **The technology was:** "SALT Agreements Signed," History.com, accessed April 21, 2021, https://www.history.com/this-day-in-history/salt-agreements-signed

134   **And verification efforts:** Bernard Gwertzman, "US Says Soviet Violates ABM Treaty," *New York Times*, February 2, 1985, https://www.nytimes.com/1985/02/02/world/us-says-soviet -violates-abm-treaty.html

134   **At the peak:** "LGM-118A Peacekeeper," Federation of American Scientists, accessed April 23, 2021, https://fas.org/nuke/guide/usa/icbm/lgm-118.htm

134   **The solution, as developed:** "Safeguard Development History," *The Military Standard*, accessed April 23, 2021, http://www.themilitarystandard.com/missile/safeguard /development.php

135   **In order to:** "What Is a Phased Array Antenna?" *everythingRF*, June 27, 2020, https://www .everythingrf.com/community/what-is-phased-array-antenna

CHAPTER 17: HUNTING BIG GAME

141   **The only difference:** "The difference between men and boys is the price of their toys," QuoteFancy, https://quotefancy.com/quote/1190836/Malcolm-S-Forbes-The-difference -between-men-and-boys-is-the-price-of-their-toys

142   **In 2009, Huawei hired:** "Huawei Hires Former BT Exec As CTO," *Mobile World Live*, October 8, 2009, https://www.mobileworldlive.com/latest-stories/huawei-hires-former-bt-exec -as-cto

142   **Controversy had dogged:** Eugenie Larson, "Bross Rides the ONI Gravy Train," *LightReading*, February 21, 2002, https://www.lightreading.com/bross-rides-the-oni-gravy-train/d/d -id/578422

143   **But Sprint was:** "Market Share Held by Mobile Cellular Services in the U.S. from 2000 to 2011, by Company," Statista, accessed April 23, 2021, https://www.statista.com /statistics/214174/us-market-share-of-mobile-cellular-services-since-2000-by-company/

143   **Sprint's stock plunged:** "Sprint Posts Huge Loss, Scraps Dividends," CNBC, February 28, 2008, https://www.cnbc.com/2008/02/28/sprint-posts-huge-loss-scraps-dividends.html

144   **Their CEO announced:** Phil Goldstein, "Huawei Taps New Telecom Firm Led by Former Sprint Exec," *Fierce Wireless*, August 24, 2010, https://www.fiercewireless.com/wireless /huawei-taps-new-telecom-firm-led-by-former-sprint-exec

144   **For years, Nortel's operations:** "TIMELINE: Key Dates in the History of Nortel," Reuters, January 14, 2009, https://www.reuters.com/article/us-nortel-timeline-sb/timeline -key-dates-in-the-history-of-nortel-idUSTRE50D3NI20090115

144   **The CEO he brought:** "Sprint Considers LTE," *Fierce Wireless*, May 23, 2010, https://www .fiercewireless.com/tech/sprint-considers-lte

145   **Amerilink secured:** Spencer E. Ante and Shayndi Raice, "Dignitaries Come on Board to Ease Huawei Into U.S.," *Wall Street Journal*, September 21, 2010, https://www.wsj.com/articles /SB10001424052748704416904575501892440266992; Stephanie Kirchgaessner, "Former US official joins Huawei consultancy," *Financial Times*, October 20, 2010, https://www.ft.com /content/cbdf6c38-dc97-11df-84f5-00144feabdc0

146   **This assertion:** Lobbying Registration Form, Senate.gov, accessed April 23, 2021, https://soprweb.senate.gov/index.cfm?event=getFilingDetails&filingID=42DCBF70 -A497-4192-AA06-E598CB19A30D&filingTypeID=1

147   **These determinations:** Daniele Lepido, "Vodafone Found Hidden Backdoors in Huawei Equipment," *Bloomberg*, April 30, 2019, https://www.bloomberg.com/news/articles/2019 -04-30/vodafone-found-hidden-backdoors-in-huawei-equipment

CHAPTER 18: LEAPFROG

148   **The threat of:** Doina Chiacu and Humeyra Pamuk, "Pompeo: 'No Mixed Messages' from U.S. on Huawei," Reuters, August 20, 2019, https://www.reuters.com/article/us-usa-china -huawei-tech/pompeo-no-mixed-messages-from-u-s-on-huawei-idUSKCN1VA1C4

148   **Hesse's strategy:** Marguerite Reardon, "The Sprint Nightmare Is Far from Over," CNet, February 28, 2008, https://www.cnet.com/news/the-sprint-nightmare-is-far-from-over/

148   **This meant he:** Phil Goldstein, "Analyst: Sprint's Network Modernization Project Could Save $2B Annually," *Fierce Wireless*, December 1, 2010, https://www.fiercewireless.com /wireless/analyst-sprint-s-network-modernization-project-could-save-2b-annually

149   **They had been:** Andy Greenberg, "The Deal That Could Have Saved Nortel," *Forbes*, January 14, 2009, https://www.forbes.com/2009/01/14/nortel-huawei-china-tech-wire-cx_ag _0114nortel.html?sh=204f6564564f

149   **they came close:** Doug Young, "Huawei-Motorola Rumors Look Logical," *Forbes*, April 13, 2012, https://www.forbes.com/sites/techonomy/2012/04/13/huawei-motorola-rumors-look -logical/?sh=41f511ad5a86

149   **Other acquisitions:** "Huawei Backs Away from 3Leaf Acquisition," Reuters, February 19, 2011, https://www.reuters.com/article/us-huawei-3leaf/huawei-backs-away-from-3leaf -acquisition-idUSTRE71I38920110219

CHAPTER 19: SNIFFING AROUND THE NUCLEAR MISSILES

158   **The provider:** Statement of commissioner Jessica Rosenworcel, FCC, accessed April 23, 2021, https://docs.fcc.gov/public/attachments/FCC-19-121A5.pdf

CHAPTER 20: EVERYONE HAS A PRICE

160   **Huawei is a symbol:** Jeanne Whalen and Yuan Wang, "Huawei Executive Becomes Unlikely Social Media Star as Chinese Rally to Tech Giant's Defense," *Washington Post*, June 12, 2019, https://www.washingtonpost.com/business/2019/06/12/huawei-executive -becomes-unlikely-social-media-star-chinese-rally-tech-giants-defense/

161   **In October of 2010:** Spencer E. Ante and Shayndi Raice, "Security Concerns Over China Stretch to ZTE," *Wall Street Journal*, October 30, 2010, https://www.wsj.com/articles/SB100 01424052702304879604575582533473050868

162   **In a spirited defense:** Daniel Lippman and Steven Overly, "China's ZTE Taps Joe Lieberman for D.C. Damage Control," *Politico*, December 13, 2018, https://www.politico.com /story/2018/12/13/zte-china-joe-lieberman-1031383

162   **If the FCC:** "Andy Purdy," The Open Group, accessed April 23, 2021, https://www.open group.org/member/andy-purdy

163   **He described:** "Samir Jain, Director of Policy," Center for Democracy and Technology, accessed April 23, 2021, https://cdt.org/staff/samir-jain/

164   **This is not good:** Saagar Enjeti, "Obama Cybersecurity Official: It's 'A Problem' That Former Colleague Now Lobbies for Huawei," *The Hill*, May 20, 2019, https://thehill.com/hilltv /rising/444566-obama-cybersecurity-official-its-a-problem-that-former-colleagues -lobbying-for

0

164　**In June of 2019:** Jeanne Whalen and Yuan Wang, "Huawei Executive Becomes Unlikely Social Media Star as Chinese Rally to Tech Giant's Defense," *Washington Post*, June 12, 2019, https://www.washingtonpost.com/business/2019/06/12/huawei-executive-becomes-unlikely-social-media-star-chinese-rally-tech-giants-defense/

164　**There are no laws:** "Huawei Denies Being Bound by Chinese Spy Laws," AFP, October 6, 2019, https://www.france24.com/en/20190610-huawei-denies-being-bound-chinese-spy-laws

165　**Article 14:** Murray Scot Tanner, "Beijing's New National Intelligence Law: From Defense to Offense," *Lawfare* (blog), July 20, 2017, https://www.lawfareblog.com/beijings-new-national-intelligence-law-defense-offense

165　**Sir Mike used:** Paul Lipscombe, "Restricting Huawei Equipment Will Hold UK Back, Warns Sir Mike Rake," *mobilenews*, March 10, 2020, https://www.mobilenewscwp.co.uk/News/article/restricting-huawei-equipment-will-hold-uk-back-warns-sir-mike-rake

165　**The list of prominent:** Richard Pendlebury and Larissa Brown, "Huawei's Very British Coup: Leviathans of the British Establishment Are Labelled China's 'Useful Idiots' for Being on the Phone Giant's Payroll . . . So What DOES Huawei Get for Its Money?" *Daily Mail*, July 7, 2020, https://www.dailymail.co.uk/news/article-8500033/Leviathans-British-establishment-labelled-Chinas-useful-idiots.html

165　**On the Chinese:** Nicola Slawson, "David Cameron to Lead £750m UK-China Investment Initiative," *Guardian*, December 16, 2017, https://www.theguardian.com/politics/2017/dec/16/david-cameron-to-lead-750m-uk-china-investment-initiative

CHAPTER 21: THE DEBT TRAP

168　**If you owe:** "Owe Your Banker £1,000 and You Are at His Mercy; Owe Him £1 Million and the Position Is Reversed," Quote Investigator, https://quoteinvestigator.com/2019/04/23/bank/

171　**China began to use:** Angus Grigg, "Debt-Trap Diplomacy: PNG Wants Huawei Loan Cancelled," *Financial Review,* August 12, 2020, https://www.afr.com/companies/telecommunications/debt-trap-diplomacy-png-wants-huawei-loan-cancelled-20200811-p55kmr

171　**With a total:** "Ethiopia: The Case for Partial Privatization of Ethio Telecom," *Africa Report*, September 1, 2020, https://www.theafricareport.com/39864/ethiopia-the-case-for-partial-privatization-of-ethio-telecom/

CHAPTER 22: HACKING A CONTINENT

173　**He described how:** Sun Tzu, "The Annotated Art of War (Parts 13.1-3: War Is Expensive)," Changing Minds, http://changingminds.org/disciplines/warfare/art_war/sun_tzu_13-1.htm

174　**He praised the:** Yara Bayoumy, "Glitzy New AU Headquarters a Symbol of China-Africa Ties," Reuters, January 29, 2012, https://www.reuters.com/article/ozatp-africa-china-20120129-idAFJOE80S00K20120129

175　**Sources told:** Ghalia Kadiri and Joan Tilouine, "A Addis-Abeba, le Siège de l'Union Africaine Espionné par Pékin," *Le Monde*, January 27, 2018, https://www.lemonde.fr/afrique/article/2018/01/26/a-addis-abeba-le-siege-de-l-union-africaine-espionne-par-les-chinois_5247521_3212.html

175 **According to investigations:** John Aglionby, Emily Feng, and Yuan Yang, "African Union Accuses China of Hacking Headquarters," *Financial Times,* January 29, 2018, https://www .ft.com/content/c26a9214-04f2-11e8-9650-9c0ad2d7c5b5

176 **She found contracts:** Danielle Cave, "The African Union Headquarters Hack and Australia's 5G Network," *The Strategist,* July 13, 2018, https://www.aspistrategist.org.au /the-african-union-headquarters-hack-and-australias-5g-network/

177 **What Huawei supplied:** Karishma Vaswani, "Huawei: The Story of a Controversial Company," *BBC,* March 6, 2019, https://www.bbc.co.uk/news/resources/idt-sh/Huawei

178 **Today, if you looked:** "Chinese Workers in Africa—What's the Real Story?" *Development Reimagined,* October 8, 2020, https://developmentreimagined.com/2020/10/08 /chinese-workers-in-africa-whats-the-real-story/

178 **The breach was:** Raphael Satter, "Exclusive-Suspected Chinese Hackers Stole Camera Footage from African Union—Memo," Reuters, December 16, 2020, https://www.reuters.com /article/us-ethiopia-african-union-cyber-exclusiv/exclusive-suspected-chinese-hackers -stole-camera-footage-from-african-union-memo-idUSKBN28Q1DB

CHAPTER 23: BUG IN THE WALLS

179 **There are just:** "The Origin of the Quote 'There Are Two Types of Companies'," *TaoSecurity Blog,* https://taosecurity.blogspot.com/2018/12/the-origin-of-quote-there-are-two-types .html

180 **Germany and France:** "Economic Intelligence Collection Directed Against the United States," Federation of American Scientists, accessed April 24, 2021, https://fas.org/irp/nsa /ioss/threat96/part05.htm

181 **Shield's persistence:** Sam Cooper, "Inside the Chinese Military Attack on Nortel," *Global News,* August 25, 2020, https://globalnews.ca/news/7275588/inside-the-chinese -military-attack-on-nortel/

183 **And, according to Shields:** Natalie Obiko Pearson, "Did a Chinese Hack Kill Canada's Greatest Tech Company?" *Financial Review,* July 10, 2020, https://www.afr.com/technology /did-a-chinese-hack-kill-canada-s-greatest-tech-company-20200706-p559gu

184 **To this day:** Tom Blackwell, "Exclusive: Did Huawei Bring Down Nortel? Corporate Espionage, Theft, and the Parallel Rise and Fall of Two Telecom Giants," *National Post,* February 24, 2020, https://nationalpost.com/news/exclusive-did-huawei-bring-down-nortel -corporate-espionage-theft-and-the-parallel-rise-and-fall-of-two-telecom-giants

184 **After a series:** David Pugliese, "The Mystery of the Listening Devices at DND's Nortel Campus," *Ottawa Citizen,* October 18, 2016, https://ottawacitizen.com/news/national /defence-watch/the-mystery-of-the-listening-devices-at-dnds-nortel-campus

CHAPTER 24: NOT IF IT'S WAR

189 **Surge forward:** Dan Strumpf, "Huawei Founder Ren Zhengfei Takes Off the Gloves in Fight Against U.S.," *Wall Street Journal,* June 6, 2020, https://www.wsj.com/articles/huawei -founder-ren-zhengfei-takes-off-the-gloves-in-fight-against-u-s-11591416028?mod=search results_pos1&page=1

192 **Huawei has responded:** Dan Strumpf, "Huawei Founder Ren Zhengfei Takes Off the Gloves in Fight Against U.S.," *Wall Street Journal,* June 6, 2020, https://www.wsj.com/articles/huawei -founder-ren-zhengfei-takes-off-the-gloves-in-fight-against-u-s-11591416028?page=1

192   **Fringe Theories:** "Former CDC Director Believes Virus Came from Lab in China," *CNN*, March 26, 2021, https://www.cnn.com/videos/health/2021/03/26/sanjay-gupta-exclusive-robert-redfield-coronavirus-opinion-origin-sot-intv-newday-vpx.cnn

193   **While the true:** Tom Mitchell, Sun Yu, Xinning Liu, and Michael Peel, "China and Covid-19: What Went Wrong in Wuhan?" *Financial Times*, October 17, 2020, https://www.ft.com/content/82574e3d-1633-48ad-8afb-71ebb3fe3dee

193   **When Australian Prime Minister:** Nick Perry, "Australian Leader Calls China's Graphic Tweet 'Repugnant'," AP, November 30, 2020, https://apnews.com/article/scott-morrison-australia-china-coronavirus-pandemic-new-zealand-8b7679f46aff690a198c220da554ab4b

193   **This provocation:** Transcript, Prime Minister's Press Conference, November 30, 2020, https://www.pm.gov.au/media/virtual-press-conference-2

CHAPTER 25: CHINA'S USE OF TECH TO OPPRESS

195   **They have started:** "China Due to Introduce Face Scans for Mobile Users," *BBC*, December 1, 2019, https://www.bbc.com/news/world-asia-china-50587098

196   **Nearly a million:** "Up to One Million Detained in China's Mass 'Re-education' Drive," Amnesty International, accessed April 24, 2021, https://www.amnesty.org/en/latest/news/2018/09/china-up-to-one-million-detained/

196   **A study released:** "Patenting Uyghur Tracking—Huawei, Megvii, More," IPVM, January 12, 2021, https://ipvm.com/reports/patents-uyghur

196   **If Beijing was hoping:** Margaret Brennan, Christina Ruffini, and Camilla Schick, "With China's Treatment of Muslim Uighurs Determined to Be Genocide, Biden Administration Under Pressure to Act," *CBS News*, January 27, 2021, https://www.cbsnews.com/news/china-treatment-of-muslim-uighurs-determined-to-be-genocide-biden-administration-under-pressure-to-act/

196   **Huawei responded:** Arjun Kharpal, "China's Huawei Tested A.I. Software That Could Identify Uighur Muslims and Alert Police, Report Says," December 9, 2020, https://www.cnbc.com/2020/12/09/chinas-huawei-tested-ai-software-that-could-identify-uighurs-report.html

197   **Barbara Boxer:** Lachlan Markay, "Scoop: Boxer to Drop Representation of Chinese Surveillance Firm," Axios, January 12, 2021, https://www.axios.com/scoop-boxer-to-drop-representation-of-chinese-surveillance-firm-07c814f9-97d9-419b-bd95-710ead26fd39.html

197   **This crackdown explicitly:** "Full Text of the Constitution and the Basic Law: Chapter II Relationship between the Central Authorities and the Hong Kong Special Administrative Region," The Basic Law of the Hong Kong Special Administrative Region of the People's Republic of China, https://www.basiclaw.gov.hk/text/en/basiclawtext/chapter_2.html

197   **After several Hong Kong:** Helier Cheung and Roland Hughes, "Why Are There Protests in Hong Kong? All the Context You Need," *BBC News*, May 21, 2020, https://www.bbc.com/news/world-asia-china-48607723

198   **Those protesters sought:** Zeynep Tufekci, "In Hong Kong, Which Side Is Technology On?" *Wired*, October 22, 2019, https://www.wired.com/story/hong-kong-protests-digital-technology/

198   **This technology:** Paul Mozur and Lin Qiqing, "Hong Kong Takes Symbolic Stand Against China's High-Tech Controls," *New York Times*, October 3, 2019, https://www.nytimes.com/2019/10/03/technology/hong-kong-china-tech-surveillance.html

198 **That's the claim:** Joe Parkingson, Nicholas Bariyo, and Josh Chin, "Huawei Technicians Helped African Governments Spy on Political Opponents," *Wall Street Journal*, August 15, 2019, https://www.wsj.com/articles/huawei-technicians-helped-african-governments-spy-on-political-opponents-11565793017

199 **His name and music:** "'Sensitise to sanitise': Bobi Wine Uses Song to Fight Coronavirus Across Africa," *Guardian*, March 26, 2020, https://www.theguardian.com/global-development/2020/mar/26/sensitise-to-sanitise-bobi-wine-uses-song-to-fight-coronavirus-across-africa

199 **According to the Times:** Jane Flanagan, "Uganda MP Robert Kyagulanyi Who Spoke Out Against President Museveni 'Tortured' Before Military Court Hearing," *Times*, August 17, 2018, https://www.thetimes.co.uk/article/uganda-mp-robert-kyagulanyi-who-spoke-out-against-president-museveni-tortured-before-military-court-hearing-tclx76tdz

200 **Not so in Uganda:** "Bobi Wine Charged with 'Annoying' Uganda's Museveni," *BBC News*, August 6, 2019, https://www.bbc.com/news/world-africa-49247860

201 **That's the allegation:** U.S. Department of Justice, Superseding indictment, January 24, 2019, https://www.justice.gov/opa/press-release/file/1125021/download

CHAPTER 26: NO WAY OUT?

204 **There's no question:** Isobel Asher Hamilton, "Former Google CEO Eric Schmidt Says There's 'No Question' Huawei Endangered US National Security," *Business Insider*, June 18, 2020, https://www.businessinsider.com/eric-schmidt-no-question-huawei-facilitated-spying-2020-6

204 **In the United Kingdom:** Ryan Browne, "British Mobile Carriers Warn Removing Huawei Will Cause 'Blackouts' and Cost Billions," CNBC, July 9, 2020, https://www.cnbc.com/2020/07/09/vodafone-and-bt-warn-about-cost-disruption-of-removing-huawei.html

205 **The stern letter:** Harry Baldock, "A Troubled Year Sees Sir Michael Rake Set to Leave Huawei UK Board," *Total Telecom*, February 15, 2021, https://www.totaltele.com/508643/A-troubled-year-sees-Sir-Michael-Rake-set-to-leave-Huawei-UK-board

206 **By early summer:** Klint Finley, "Huawei Still Has Friends in Europe, Despite US Warnings," *Wired*, April 25, 2019, https://www.wired.com/story/huawei-friends-europe-despite-us-warnings/

206 **In his 1990 book:** Daniel Yergin, *The Prize* (New York: Free Press, 2008), xv

207 **One company after:** John Xie, "Huawei's Survival at Stake as US Sanctions Loom," VOA News, September 14, 2020, https://www.voanews.com/silicon-valley-technology/huaweis-survival-stake-us-sanctions-loom

207 **Indeed, without access:** C. Scott Brown, "The Huawei ban explained: A complete timeline and everything you need to know," *Android Authority*, February 7, 2021, https://www.androidauthority.com/huawei-google-android-ban-988382/

207 **At the company's:** "Huawei: 'Survival Is the Goal' As It Stockpiles Chips," *BBC News*, September 23, 2020, https://www.bbc.com/news/technology-54266531

207 **Each state-of-the-art:** Dean Takahashi, "Globalfoundries: Next-Generation Chip Factories Will Cost at Least $10 billion," *VentureBeat*, October 1, 2017, https://venturebeat.com/2017/10/01/globalfoundries-next-generation-chip-factories-will-cost-at-least-10-billion/

207 **Not for lack:** Cyrus Lee, "Thousands of Taiwanese Chip Experts Moved to China for Better Pay: Reports," ZDNet, December 5, 2019, https://www.zdnet.com/article/thousands-of-taiwanese-chip-experts-moved-to-china-for-better-pay-reports/

207 **This program is off:** Mark Lapedus, "China Speeds Up Advanced Chip Development," *Semiconductor Engineering,* June 22, 2020, https://semiengineering.com/china-speeds-up -advanced-chip-development/

207 **The results:** Joe Panettieri, "Huawei: Banned and Permitted in Which Countries?" *Channel e2e,* March 11, 2021, https://www.channele2e.com/business/enterprise/huawei -banned-in-which-countries/2/

208 **Ericsson, restricted as:** Monica Alleven, "Ericsson CEO Lobbied Swedish Minister over Huawei Ban—Report," *Fierce Wireless,* January 4, 2021, https://www.fiercewireless.com /wireless/ericsson-ceo-lobbied-swedish-minister-over-huawei-ban-report

208 **Chancellor Angela Merkel:** Katrin Bennhold and Jack Ewing, "In Huawei Battle, China Threatens Germany 'Where It Hurts': Automakers," *New York Times,* January 16, 2020, https://www.nytimes.com/2020/01/16/world/europe/huawei-germany-china-5g-auto makers.html

CHAPTER 27: 5G IS DIFFERENT

210 **5G will be:** Sadanand Dhume, "India Will Have a Say in Whether China Dominates 5G," *Wall Street Journal,* October 31, 2019, https://www.wsj.com/articles/india-will-have -a-say-in-whether-china-dominates-5g-11572561367

CHAPTER 28: DANGEROUS DESIGNS: WHY WE NEED TO CARE

216 **A clever tech geek:** Alex Schiffer, "How a Fish Tank Helped Hack a Casino," *Washington Post,* July 21, 2017, https://www.washingtonpost.com/news/innovations/wp/2017/07/21 /how-a-fish-tank-helped-hack-a-casino/

219 **The results of such a collapse:** Erin Douglas, "Texas was 'seconds and minutes' away from catastrophic monthslong blackouts, officials say," *Texas Tribune,* February 18, 2021, https:// www.texastribune.org/2021/02/18/texas-power-outages-ercot/

219 **Increasingly, such information:** "The Leading Remote Monitoring and Manage-ment Tools for Utilities," DP Stele, accessed April 25, 2021, https://www.dpstele.com /insights/2019/11/14/remote-monitoring-and-management-tools-utilities/

222 **His flamboyant prank:** Paul Marks, "Dot-Dash-Diss: The Gentleman Hacker's 1903 Lulz," *New Scientist,* December 20, 2011, https://www.newscientist.com/article/mg2122 8440-700-dot-dash-diss-the-gentleman-hackers-1903-lulz/

CHAPTER 29: IF IT'S SO RISKY . . .

223 **If we do this right:** "Remarks of Commissioner Jessica Rosenworcel," FCC, September 14, 2020, https://docs.fcc.gov/public/attachments/DOC-366876A1.pdf

224 **Companies have already rolled:** Bill Siwicki, "A Guide to Connected Health Device and Remote Patient Monitoring Vendors," *Healthcare IT News,* May 6, 2020, https://www.health careitnews.com/news/guide-connected-health-device-and-remote-patient-monitoring -vendors

224 **Soon, 5G wireless:** "Cardiovascular Treatments & Therapies," Abbott, accessed April
25, 2021, https://www.cardiovascular.abbott/us/en/patients/living-with-your-device
/arrhythmias/remote-monitoring.html; Albert Cai, "Insulin Delivery Systems Coming in
2020," *Diatribe*, December 2, 2019, https://diatribe.org/tech-horizon-automated-insulin
-delivery-systems-coming-2020

225 **The basic concept:** Scott Gottlieb, "Surgeons Perform Transatlantic Operation Using
Fibreoptics," *The BMJ*, September 29, 2001, https://www.ncbi.nlm.nih.gov/pmc/articles
/PMC1121281/

225 **In March of 2019:** "China Performs First 5G-Based Remote Surgery on Human Brain," *China
Daily*, March 18, 2019, http://www.chinadaily.com.cn/a/201903/18/WS5c8f0528a3106c
65c34ef2b6.html

CHAPTER 30: BUILDING A RESPONSE

231 **When disruptive innovations:** "Clayton M. Christensen: An Interview by Bob Morris,"
Blogging on Business, June 9, 2011, https://bobmorris.biz/clayton-b-christensen-a-book
-review-by-bob-morris

234 **The European vendors:** Cheetan Woodun, "Ericsson and Nokia: Assessing Geopolitical Risks," *Seeking Alpha*, January 14, 2021, https://seekingalpha.com/article/4398942
-ericsson-and-nokia-assessing-geopolitical-risks

236 **But the proof:** Mark Lapedus, "Semiconductor R&D Crisis?" *Semiconductor Engineering*,
June 11, 2015, https://semiengineering.com/semiconductor-rd-crisis/

CHAPTER 31: OUR SUPERPOWER

237 **We cannot regulate:** David Paul, "What Has Happened to Chinese Business Mogul Jack Ma?"
Digit, January 6, 2021, https://digit.fyi/what-happened-to-jack-ma-chinese-billionaire/

240 **For example, Herb Kelleher:** Keli Flynn, "Southwest Airlines," Texas State Historical Association, accessed April 25, 2021, https://www.tshaonline.org/handbook/entries
/southwest-airlines

241 **He knew that the:** Arjun Kharpal, "Apple vs FBI: All You Need to Know," CNBC, March 29,
2016, https://www.cnbc.com/2016/03/29/apple-vs-fbi-all-you-need-to-know.html

241 **Not so for Chinese:** Chris Buckley, "China's 'Big Cannon' Blasted Xi. Now He's Been Jailed
for 18 Years," *New York Times*, September 22, 2020, https://www.nytimes.com/2020/09/22
/world/asia/china-ren-zhiqiang-tycoon.html

242 **Jack Ma, the richest man:** "Chinese Billionaire Jack Ma Makes First Public Appearance
in Three Months," Reuters, January 20, 2021, https://www.nbcnews.com/news/world
/chinese-billionaire-jack-ma-makes-first-public-appearance-three-months-n1254845

242 **Such talk is:** Rob Davies and Helen Davidson, "The Strange Case of Jack Ma and His Three-
Month Vanishing Act," *Taipei Times*, January 27, 2021, https://www.taipeitimes.com/News
/editorials/archives/2021/01/27/2003751277

244 **But today more than:** Ian Hathaway, "Almost Half of Fortune 500 Companies Were
Founded by American Immigrants or Their Children," Brookings (blog), December 4,
2017, https://www.brookings.edu/blog/the-avenue/2017/12/04/almost-half-of-fortune-500
-companies-were-founded-by-american-immigrants-or-their-children/

244 **The country's culture:** Mindy Wright, "Top 75 CEOs of China, 2018," *CEO World*, October 16, 2018, https://ceoworld.biz/2018/10/16/top-ceos-of-china-2018/

CHAPTER 32: THE ROLE OF GOVERNMENT

245 **One of the things:** Elizabeth Redden, "The Chinese Student Threat?" *Insider Higher Ed*, February 15, 2018, https://www.insidehighered.com/news/2018/02/15/fbi-director-testifies -chinese-students-and-intelligence-threats

247 **Although new entrants:** Mike DeBonis, "Uber Car Service Runs Afoul of D.C. Taxi Commission," *Washington Post,* January 11, 2012, https://www.washingtonpost.com/blogs /mike-debonis/post/uber-car-service-runs-afoul-of-dc-taxi-commission/2012/01/11 /glQAxH3UrP_blog.html

250 **The Cyberspace Solarium:** Cyberspace Solarium Commission, March 2020, https://www .solarium.gov/

251 **A 2018 report from:** "2017 Special 301 Report," Office of the United States Trade Representative, April 2017, https://ustr.gov/sites/default/files/301/2017%20Special%20301%20 Report%20FlNAL.PDF

252 **New entrants were:** Kif Leswing, "Companies Have Bid $81 Billion for the Airwaves to Build 5G, and Winners Will Be Revealed Soon," CNBC, January 31, 2021, https://www.cnbc .com/2021/01/31/5g-spectrum-auction-bids-total-80point9-billion-winners-revealed-soon .html

CHAPTER 33: RIDING THE WAVE OF CHAOS

254 **I've searched all the parks:** "Gilbert K. Chesterton Quotes," BrainyQuote, https://www .brainyquote.com/quotes/gilbert_k_chesterton_156938

CHAPTER 34: THE NEW ECOSYSTEM

261 **When one door closes:** "When One Door Closes Another Opens, But Often We Look So Long Upon the Closed Door That We Do Not See the Open Door," Quote Investigator, https://quoteinvestigator.com/2018/12/03/open-door/

263 **Some, like Tech Mahindra:** "IBM, Tech Mahindra Collaborate to Create USD 1 Billion Ecosystem in 3 Years," *Financial Express*, February 16, 2021, https://www.financialexpress .com/industry/ibm-tech-mahindra-collaborate-to-create-usd-1billion-ecosystem-in-3 -years/2196073/

267 **Stephen Bye, an executive:** "Forum on 5G Open Radio Access Networks," Federal Communications Commission, September 14, 2020, https://www.fcc.gov/news-events/events /forum-5g-virtual-radio-access-networks

INDEX

ABOUT THE AUTHOR

 fter an early career as a writer and mar-
keter on Madison Avenue, Jon Pelson
spent nearly twenty-five years working as an
executive at some of the world's largest telecom equipment makers and service providers,
helping create and market wireless products
and solutions. During this time, he traveled
to China and watched that country's fledgling
telecommunications companies grow and
eventually seize the world lead.

Pelson has a degree in economics from
Dartmouth College and an MBA from the Darden School at the University
of Virginia. He lives with his family in Great Falls, Virginia.